An Early History of

Recursive Functions
and Computability

from Gödel to Turing

ROD ADAMS

Docent

Press

DOCENT PRESS
Boston, Massachusetts, USA
www.docentpress.com

Docent Press publishes monographs and translations in the history of mathematics for thoughtful reading by professionals, amateurs and the public.

Cover design by Brenda Riddell, Graphic Details, Portsmouth, New Hampshire.

Publisher's Preface

In researching the early history of recursive function theory for his PhD thesis, Rod Adams corresponded with some of the individuals who wrote that history including Stephen Kleene, Alonzo Church, J. Barkley Rosser and Hao Wang. The results of this correspondence were included in the thesis but the source documents, the letters from these individuals answering Adams' questions, were not. This republication of Adams' thesis has offered the opportunity to make these previously unpublished letters part of the historical record.

The body of this republication has been typeset directly from the original typewritten manuscript of the author's 1983 Doctor of Philosophy thesis, "A History of the Theory of Recursive Functions and Computability with Special Reference to the Developments Initiated by Gödel's Incompleteness Theorems," which was written in the School of Information Sciences at The Hatfield Polytechnic under the supervision of Dr. Dale Johnson.

Dr. Adams has overseen the editing and production of this new edition of his thesis. A small number of typographic errors in the original thesis have been corrected, some additions have been made to clarify the text and additional details added to a few of the bibliographic references. The source documents referred to in the first paragraph above have been included in a second appendix.

Dr. Adams is currently a Professor of Neural Computation at the University of Hertfordshire.

One of the objectives of the republication of the thesis is to present the material in a typographic format that is familiar to the modern reader. To this end, the thesis including all mathematical and symbolic logic notation has been rendered in the TEX mathematical typesetting language following the guidelines of the American Mathematical Society.

Scott Guthery, Publisher
Docent Press, LLC
Boston, Massachusetts
www.docentpress.com

May, 2011

Abstract

There are many problems of elementary number theory in which it is required to find an effectively calculable function of positive integers satisfying certain conditions, as well as a large number of problems in other fields which are known to be reducible to problems of this type in number theory. Church's thesis, published in 1936, proposed that the class of general recursive functions and the class of λ-definable functions should be equated with this intuitive class of effectively calculable functions.

This book traces the development of recursive functions from their origins in the late nineteenth century, where recursion was first used as a method of defining simple arithmetic functions, up to the mid-1930's, when the class of general recursive functions was introduced by Gödel, developed on the formal side by Kleene and utilized by Church in his thesis.

Skolem's and Hilbert's use of recursive functions as constructive, finitely defined functions are studied. Gödel's significant use of recursive functions in his proof of the incompleteness theorems is discussed, including his statement of the first precise definition of the class of functions involved.

The interactions between Church, Kleene and Rosser, who developed the notion of λ-definable functions in the early 1930's are considered in detail. The book explains how the proposal given in Church's 1936 paper, now known as Church's thesis, first arose. The influence of Gödel on Church, which led to Gödel's subsequent introduction of the general recursive functions, is examined. The development of Kleene's normal form theorem is analyzed and its significance is shown.

Finally the book briefly considers another class of functions, the Turing computable functions, that were specifically created to be equivalent to the class of effectively calculable functions. Indeed, many mathematicians find them to be the most convincing argument in favor of Church's thesis.

Rod Adams
September, 1983

Acknowledgements in the Thesis

I would like to express my gratitude to Dr. Dale Johnson, of The Hatfield Polytechnic, and Dr. G. T. Kneebone, of Bedford College, London, my supervisors, for all their encouragement, help and guidance throughout this project. I would also like to thank Professor S. C. Kleene, of the University of Wisconsin, for reading the whole of an early draft of this thesis and for making numerous valuable suggestions.

Thanks also go to Mary Callan for typing the original thesis.

Special appreciation is acknowledged for the patient encouragement and help offered by my wife, Liz.

Rod Adams
September, 1983

Contents

CHAPTER 1

Early Recursive Definitions

1.1. Introduction

This book traces the development of recursive function theory from its origins in the late nineteenth century until its maturity in the 1930's. By this later date it had become a branch of mathematics comparable in importance to group theory or to projective geometry. The emphasis of the book will be on the work of the 1930's when recursive function theory came to the notice of the mathematical world at large thanks to the work of Gödel, Church, Kleene, Rosser and Turing.

The definition of a function by recursion is the analogue for introducing functions of the method of proof by induction. At its simplest level a number-theoretic function, $f(n)$, is said to be recursively (or inductively) defined if an explicit definition of $f(0)$ is given and if $f(n + 1)$ is defined in terms of $f(n)$ by means of functions that are already available. That is to say that a method is given to determine $f(0)$, $f(1)$, $f(2)$, ... and so on in succession, so that the function is defined for all positive integers. Although this method of definition of a function does not allow one to eliminate the defined operation, as would be the case in the more usual explicit definition, it does allow the whole range of values of the new function to be unambiguously produced from previously calculated values of the function by some given iterative process.

The recursive method of defining functions is closely connected with the method of mathematical induction, a process that can be seen implicitly in the work of Euclid, and which, according to Morris Kline in his *Mathematical Thought from Ancient to Modern Times*, was recognized explicitly by Maurolycus in 1575. However, the first use of this induction process to define functions and the first time the process was actually named did not occur until the latter half of the nineteenth century. This occurrence was in the works of such mathematicians as Grassmann, Peirce, Dedekind and Peano, where the use of recursive definitions of operations on numbers was used as an aid to the understanding of the nature and properties of these numbers. Since this represents the start of recursive function theory, we will begin the book with a consideration of this area of work.

The next important stages in the evolution of recursive function theory came with Skolem's use of recursive definitions as a primary element in his foundations of arithmetic, and with Hilbert's adoption of them as an essential part of his finitary mathematics used in his metamathematical approach to the proof of the consistency of arithmetic. A discussion of these developments - will be found in Chapters 2 and 3. A consideration of the motivation behind Skolem's and Hilbert's study of the foundations of arithmetic is essential to the understanding of their work. The areas of mathematical activity that motivated Skolem, Hilbert and many other mathematicians in the early part of this century to undertake this study will be briefly reviewed in this opening chapter. These areas of mathematical activity include the rise of the general axiomatic approach to mathematics in the nineteenth century and the development of non-Euclidean geometry, with the effect that this had on the concept of mathematical truth. The final and most urgent factor to motivate this study of the foundations of arithmetic was, however, the creation of infinite set theory, together with the resulting paradoxes that were almost immediately found to affect its foundations and the concern that these paradoxes might also apply to the foundations of all other branches of mathematics.

Hilbert's attempt at proving the consistency of arithmetic was making steady progress when, in 1930, Gödel announced perhaps the most outstanding result in logic in the twentieth century. He proved that, for any formal system designed to contain arithmetic, the system, if it was consistent was incomplete and furthermore there could be no proof of the consistency of arithmetic formalizable within this system.[1] These results were published in his 1931 paper [**51**]. In this paper Gödel produced the first precise definition of recursive functions and then proceeded to make substantial use of such functions in his proof of these theorems. In Chapter 3 we will consider Hilbert's and Gödel's contributions to recursive function theory in more detail.

Gödel's theorems exerted a deep influence in the field of the foundations of mathematics and in particular they led to a complete revision of Hilbert's program. Gödel's paper was an inspiration to many mathematicians not only for the results it contained but also for the methods he introduced to produce these results. Some of these methods, such as his arithmetization of the metamathematics have become everyday tools in later foundational papers. His use of recursive functions inspired much further study of this class of functions; Rózsa Péter was perhaps the most prolific author in this respect. Research was initiated into a consideration of the limitations of formal systems in general, and this culminated in a study of the effective calculation procedures used in everyday mathematical activity. In Chapter 6 we consider Churchs thesis, which is not just his PhD. thesis, but a famous proposal contained in his 1936 paper [**23**], that has since been known simply as Church's thesis and which is referred to as such in this book. This proposal states that a formal counterpart to the class of effectively calculable functions could be given and that in fact two equivalent classes of functions already existed to satisfy this need. One of these formal characterizations, λ-definable functions, arose in Princeton from the work of Church, Kleene and Rosser. In Chapter 4 we will investigate

---

[1]A more precise statement of Gödel's results must wait until Chapter 3 where his theorems are discussed in more detail.

how this class of functions developed. Another formal characterization, general recursive functions, was proposed by Gödel himself, acting on a suggestion of Herbrand. These functions, in fact, came directly from a consideration of Gödel's 1931 results and an attempt to generalize the concept of recursive functions used in that paper. We will consider the development of these functions in Chapter 5.

Finally in Chapter 7 another characterization, that of Turing computable functions, will be considered along with Turings thesis (proposed in his 1936 paper [160]) which stated that these functions were also equivalent to the effectively calculable functions

During the course of the book we will survey the problems that faced the mathematicians and logicians who were concerned with the foundations of mathematics, and indicate how the solutions, attempted solutions and indeed failed solutions, led to new and important concepts and techniques. The overall driving force behind the desire to study the foundations of mathematics was the concern that mathematics might be inconsistent. This problem is still recognized today: mathematicians have had to learn to live with the knowledge that there does not exist any universally accepted proof of the consistency of mathematics. But, despite the strenuous efforts made by many different inquirers no actual inconsistency has ever been found. Consequently confidence in mathematics is still very high despite this failure to prove its consistency once and for all. As the group of French mathematicians who write under the pseudonym of Nicolas Bourbaki have commented:

> There are now twenty-five centuries during which the mathematicians have had the practice of correcting their errors and thereby seeing their science enriched, not impoverished; this gives them the right to view the future with serenity.[2]

---

[2]See [97, p. 1210].

The remainder of this introductory chapter will consist of a description of the first uses of recursive definitions, and a brief analysis of the underlying trends in mathematics that were to influence the future course of recursive function theory.

1.2. The First Recursive Definitions

By the end of the eighteenth century mathematicians had begun to be concerned about the looseness in the concepts and proofs in analysis. It was becoming recognized that most of the fundamental concepts of analysis were not very clearly defined and consequently confusion could arise whenever these concepts were used. During the nineteenth century several mathematicians attempted to correct this state of affairs by instilling rigor into analysis.

Bolzano's detailed and careful study, undertaken in the early part of the century, was probably the first of such attempts. Unfortunately his work went unnoticed for the first half of the century and consequently, as far as the early attempts at making analysis more rigorous are concerned, Cauchy's work is probably the most well known. The next major figure in this field was Weierstrass who, in the mid-nineteenth century, improved on the work of Cauchy, Bolzano and others by continuing to build the foundations of analysis on arithmetic concepts and thereby freeing analysis from its dependence on geometry and intuition.

All of this work presupposed the soundness of the real number system. Towards the latter half of the nineteenth century it became accepted that this, too, needed to have a proper foundation. Weierstrass was perhaps the first to recognize this need, and in his lectures at Berlin University he began to produce a theory of irrationals based on rationals and a theory of the rationals based on the integers. Cantor, in 1872, published an account of how to found the irrational numbers and Dedekind, also in 1872, published another method in his influential book, *Stetigkeit and irrationale Zahlen* [**37**]. But all this work still left undone the task of giving the integers a logical foundation. Some mathematicians, in fact, believed that the integers were so fundamental that no foundation for them was possible. This position was taken by Kronecker who, in a speech on 21 September 1886, said, "God made the integers, all else is the work of man." However, as we shall see, others did believe that it was possible to give the integers a logical basis.

Also in progress during this period was a movement to improve on Aristotelian logic. Starting with work carried out by such mathematicians as De Morgan and Boole, logic was made more algebraic in nature and extended in its use. The logic of relations was developed, mainly by De Morgan and Peirce, and generally logic was recast so that it could be manipulated by using the ordinary rules of arithmetic. All this new material on logic was collected and organized by Schröder in his *Vorlesungen Über die Algebra der Logik* which was published between 1890 and 1905. Another trend, initiated by Frege in his pioneering work of 1879, was to make logic into a formal system where no intuitive notions would be needed. He believed that this formal structure could then be used to found the whole of mathematics. Frege's thorough and precise study also contained the first codification of quantification theory and can be considered to have transformed the current logic into our modern form of logic in one step. As we shall see, the development of logic and the analysis of the foundations of arithmetic became interrelated as each influenced the development of the other.

Two of the first attempts at analyzing the properties of numbers were by Hermann Grassmann in 1861. and Charles Sanders Peirce in 1881. As Peirce states in [**63**, p. 158]:

> The object of this paper is to show that [the elementary propositions concerning numbers] are strictly syllogistic consequences from a few primary propositions. The question of

the logical origin of the latter which I here regard as definitions, would require a separate discussion.

Both Peirce and Grassmann use recursive definitions when defining addition
and multiplication of natural numbers and both use mathematical induction to
prove the laws of arithmetic. The recursive definitions contained in Grassmann's
paper must therefore represent the first of such definitions, but neither Peirce
nor Grassmann named their definitional procedures and they both gave their
definitions informally. For instance, in Peirce's paper:[3]

> By $x + y$ is meant, in the case $x = 1$, the number next greater
> than y; and in other cases, the number next greater than $x' + y$,
> where x' is the number next smaller than x. By $x \times y$ is meant,
> in case $x = 1$, the number y, and in other cases $y + x'y$ where
> x' is number next smaller than x.

Hence, when it came to analyzing the basic concepts concerning operations
on natural numbers, recursive definitions seemed to Grassmann and Peirce to
be the most natural approach. Peirce's work was not, in fact, very well known
outside the United States of America and so it had little immediate impact.

The next occurrence of recursive definitions used in attempts at founding
arithmetic was in Dedekind's 1888 essay, *Was sind und was sollen die Zahlen?*.
In this essay Dedekind attempted to base the arithmetic of natural numbers
on such logical notions as class, union, intersection and mapping of one class
onto another. Using these set-theoretic concepts Dedekind proceeded to define natural numbers and, also using the primitive notions of the number one
and the successor operation, he defined recursively addition, multiplication and
powers of natural numbers. His powerful Theorem 126 supplied the method
of justifying these recursive definitions in terms of his set-theoretic notions.

---

[3]See [**63**, p. 160].

Dedekind seems to be the first person to use the term "definition by recursion" and examples of such definitions given in his essay include:[4]

Addition

$$m + 1 = m'$$
$$m + n' = (m + n)'$$

Multiplication

$$m \cdot 1 = m$$
$$m \cdot n' = m \cdot n + m$$

Exponentiation

$$a^1 = a$$
$$a^{n'} = a \cdot a^n = a^n \cdot a$$

where the symbol $'$ denotes the successor function.

These are much more modern, symbolic forms of recursive definitions than those which were given previously by other authors. Dedekind thought that these recursive definitions and proof by induction were very important parts of his system, as can be seen from the following extracts from his letter to Keferstein in which he explained the thoughts that led to his 1888 essay:

> Does what has been said so far also contain a *method of proof* sufficient to establish, in full generality, propositions that are supposed to hold for *all* numbers n? Yes! The famous method of proof by induction.[5]
>
> Finally, is it possible also to set up the *definitions* of numerical operations consistently for *all* numbers n? Yes! This is in fact accomplished by the theorem of Article 126 of my essay.[6]

Dedekind was trying to base the natural numbers and hence all of arithmetic on logic but he is not new considered to have been fully successful due to his

---

[4]See [**39**, p. 97, 101, 104].

[5]He gives a set-theoretic definition of proof by induction in Articles 59, 60 and 80.

[6]This theorem being the one that gives a set-theoretic justification of his method of definition by recursion. See also [**64**, p. 101].

use of such imprecise phrases as "abstracting altogether from the nature of the elements." His essay, though, must be regarded as a classic of its time. It is a precise and mature piece of work and he did free the concept of number from its dependence on intuition. Much of Dedekind's work has been used by later mathematicians, and Peano, in particular, acknowledged that his axioms were derived from Dedekind's work. Since Dedekind was the first to name, as well as recognize fully the power of recursive definitions, he is generally regarded as the founder of recursive function theory.

The next significant contributor to this field was Peano. Peano was not convinced that the primitive notions of arithmetic could be reduced to logic so instead he proceeded to base arithmetic on an axiomatic foundation. Peano designed a new symbolic language and then continued by developing logic and mathematics at the same time. For him logic was just a servant of mathematics. Peano also recognized the importance of recursive definitions and made substantial use of them. He made considerable use of the work of previous authors, in particular Grassmann, for his arithmetic, and Dedekind, for his work on the foundations of numbers.

What Peano did, beginning in his 1889 paper [**116**], was to base arithmetic on certain formal axioms and to improve the symbolic language of mathematics. Peano based his treatment of arithmetic on the natural structural features of everyday experience of numbers as they are used in counting and reckoning. The fact that they form a natural sequence or progression, each being obtainable in regular succession by unlimited repetition of a single generating act, led him to base his arithmetic entirely on the following undefined notions:

    i) the set of numbers
    ii) the number one
    iii) the successor operation
    iv) the property of being equal to.

He then abstracted nine axioms of which we now recognize four as belonging
to the underlying logic. The remaining five axioms are generally referred to
as the Peano axioms. The fifth of these is the principle of induction. These
axioms are very similar to the ones found in Dedekind's 1888 essay. [**37**, p. 67]
When it came to defining his arithmetic operations Peano also used recursive
definitions, these being the constructive counterpart to mathematical induction
which was his sole proof procedure. Examples of his recursive definitions are:[7]

$$a \in N \Rightarrow a \times 1 = a$$
$$a, b \in N \Rightarrow a \times (b + 1) = a \times b + a$$
$$\left.\right\} \text{Multiplication}$$

$$a \in N \Rightarrow a^1 = a$$
$$a, b \in N \Rightarrow a^{b+1} = a^b \times a$$
$$\left.\right\} \text{Exponentiation}$$

where N is the set of numbers.

As can be seen, these are very similar to Dedekind's definitions. Dedekind
could justify his recursive definitions in terms of his set theoretic notions. Peano,
on the other hand, gave no such method of justification in his paper. Peano,
therefore, should have introduced these recursive functions as extra axioms
instead of as definitions when, as van Heijenoort has pointed out,[8] they did not
even satisfy his own statement of what a definition should be, namely:

> ... where a [the right hand side of the definition] is an aggregate
> of signs having a known meaning.

Also Peano's logic had no rules of inference so he could never properly
deduce his theorems. Consequently his logic has not survived, but his symbolism, and in particular his axioms for arithmetic, have figured frequently in
subsequent literature.

---

[7]See [**64**, p. 96].
[8]See [**64**, p. 93].

In summary we see that recursive definitions were used naturally by mathematicians trying to analyze the concept of natural numbers. Whether this was for a grander plan, like Dedekind's, to found arithmetic on logic, or just to abstract formal axioms, like Peano, they were all trying to treat arithmetic rigorously.

1.3. Mathematical Thought at the Turn of the Nineteenth Century

Advances in recursive function theory were intimately related to considerations of the foundations of arithmetic. In this section we shall briefly consider some of the developments in mathematics during the nineteenth century which were to influence subsequent progress in the field of the foundations of arithmetic. These developments include the change of attitude towards mathematical truth engendered by the non-Euclidean geometries and the resultant effects on axiom systems, and also the paradoxes in set theory and logic. We shall also consider some of the early attempts at avoiding the paradoxes.

1.3.1. Mathematical Truth and Consistency. When Euclid produced his axioms for geometry in the third century BC it would have been thought absurd to consider whether or not they were contradictory since mathematics was understood to be an accurate (if idealized) description of real phenomena. Mathematics was a body of truths about the real world and the function of using mathematics was to uncover the mathematical design of the universe. From his axioms (which he thought of as absolute truths) Euclid produced, for that time, a rigorous treatment of geometry.[9]

---

[9]It would have been impossible for Euclid to axiomatize his theory fully because he would have needed at least two modern developments, that is, Frege's discovery and axiomatization of quantification theory and the Dedekind-Peano axioms of arithmetic.

Hence Euclid thought of his geometry as a true description of space whereas modern mathematicians regard Euclidean geometry as one of a number of different possible mathematical models of space. This change of outlook arose mainly because of the development of non-Euclidean geometries in the nineteenth century, these geometries being seen as just as applicable to the real world as Euclidean geometry. Hence with several different sets of geometric axioms existing, each having a valid claim to represent the real world, the truth of these sets of axioms could no longer be guaranteed by the belief that they were descriptions of the real world. Doubt was also thrown on arithmetic truth - under what conditions were the numbers and operations on them applicable to the real world? It became apparent that an axiomatic system could be constructed and studied in its own right using axioms that could be any undefined initial principles. Only later might its applicability to the real world be considered.

During the nineteenth century mathematicians strove to make arithmetic and analysis more rigorous, and the axiomatic method was the major technique used in this task. As we have seen, Peano used the axiomatic method to found the natural numbers. The axioms that he used being simple basic propositions. Once he had succeeded in axiomatizing the natural numbers he proceeded to generate in turn the integers, the rational numbers and the real numbers from this axiomatic base. Hilbert, on the other hand, thought it better to produce one set of axioms for the whole of the real number systems. He did this in his 1900 paper, "Über den Zahlbegriff" [70], in which he gave a set of eighteen simple propositions for the real numbers from which he produced all of the theorems about real numbers. He believed that this full axiomatic method for presenting the real numbers was better than the genetic one as used by Peano and others. As long as mathematics was regarded as the truth about nature the possibility that contradictory theorems could arise did not occur, but once it was recognized that the axioms were arbitrary then contradictions might occur. As Cantor observed in 1883:[10]

---

[10]See [97, p.1031].

> Mathematics is entirely free in its development and its concepts
> are restricted only by the necessity of being non-contradictory.

Thus by the end of the nineteenth century it was recognized that it was necessary to guarantee the consistency of all mathematical structures built on axioms. Hilbert, after he had produced his axioms for the real numbers, recognized this need for proving the consistency ar axioms since he stated that the real numbers would exist in a mathematical sense when the axioms were proved consistent. However he was not aware at the time of the difficulty of proving the consistency of such axioms.

Various consistency results had already been proved at this time but the proofs were all relative consistency proofs, for instance, non-Euclidean geometry was proved to be consistent providing Euclidean geometry was consistent. Hilbert himself, in his influential treatise on geometry, *Grundlagen der Geometrie* [**68**], proved that Euclidean geometry was consistent if arithmetic was consistent. But eventually it was realized that the relative consistency procedure had to be supplemented by proving the absolute consistency of at least one structure such as arithmetic. This problem of proving the consistency of arithmetic was posed by Hilbert as the second problem in his famous list of twenty-three problem for the new century, which he presented at the Second International Congress of Mathematicians in Paris in 1900. [**69**]

Hilbert stressed this problem as being the basic problem in the foundations of mathematics and he thought, at the time, that a suitable modification of the methods of Dedekind and Weierstrass in the theory of irrational numbers would suffice to obtain the desired proof of consistency. However, as we shall see in Chapter 3, by 1904 he had come round to a different way of thinking. The need for a proof of the consistency of arithmetic was seen to be much more urgent as the paradoxes in logic and set theory became widely known.

1.3.2. The Paradoxes of the Infinite. The first paradoxes appeared in transfinite set-theory, the branch of mathematics that was created virtually single-handedly by Cantor. While dealing with sets of real numbers Cantor introduced some notions about infinite sets of points and then later, realizing the importance of infinite sets, he undertook the study of infinite sets in detail.

Cantor's theory of sets was a bold and imaginative creation but one that, by its very nature, drew much criticism. The use of infinity had been the subject of contention ever since the time of the Greeks. The concept of an actual infinite number of objects had been denied existence so that the potential infinite was the only recognized use of the notion of infinity. Cantor faced the difficulties associated with infinite sets and, building on, in his terms, clear and certain ideas, he defined ordinal and cardinal numbers for sets of any size and proved many results in transfinite set theory. His use of infinite sets was bound to cause controversy and, sure enough, his ideas immediately aroused much criticism, especially from Kronecker who was very vehement in his opposition. But other mathematicians, such as Hilbert, were full of praise for Cantor's original and powerful new theory, which they thought of as a profound and beautiful product of the human intellect. Unfortunately almost immediately after it was created, paradoxes began to appear.

In 1897 Cesare Burali-Forti published the first paradox associated with this new theory of transfinite numbers [11]. Cantor had himself discovered the paradox in 1895. It can be stated as follows:

> The sequence of all ordinal numbers is well-ordered and and therefore should have the greatest of all ordinal numbers as its ordinal number. But this ordinal number would then be greater than all ordinal numbers, which would be a contradiction.

This was the first paradox but others were soon formulated, some of them being couched in very simple terms that were fundamental to the foundations of this new theory. They showed that Cantor's intuitive notions were not sufficiently rigorous to avoid contradictions.

A possible solution was to axiomatize Cantor's set theory. This approach was taken by Zermelo [177] and subsequently modified in slightly different ways by Fraenkel [42], von Neumann [110], Skolem [152] and Weyl [170],[171]. A precise characterization of what constitutes a set was given and in this way all known paradoxes were eliminated while, at the same time, the resulting axiomatic set theory was adequate for developing practically all of classical analysis. But the question of the consistency of set theory had not actually been solved: all the known paradoxes were eliminated by this axiomatization but there was no guarantee that others would not be found in the future.

In 1901 Russell discovered a paradox that had been independently discovered by Zermelo involving only the notions of set and element. The paradox which disturbed many mathematicians by its sheer simplicity fell squarely into the realm of logic. The knowledge of the existence of paradoxes in logic itself indicated that perhaps there were defects in the previously trusted methods of constructing and reasoning about mathematical objects. Also arithmetic had, by various authors, been founded on logic and on set theory, so paradoxes in both these disciplines threw doubt on the foundations of arithmetic and hence on the foundations of all of mathematics. The foundations of mathematics needed to be proved secure.

All of these paradoxes made use of self-referential statements, the set of all sets for example, and some mathematicians, such as Poincaré [124]–[125], [126] and Russell [141], [143], thought that the causes of the paradoxes might lie with this use of self-reference. Unfortunately self-reference is essential in some parts of classical analysis, for instance the notion of least upper bound is defined in terms of all upper bounds. This threw doubt on the foundations of classical analysis and, since the consistency of analysis could be reduced to that of the arithmetic of natural numbers, the question of the consistency of arithmetic consequently became a matter of much greater urgency.

Various methods for removing the danger of the paradoxes were suggested. One way, known as Logicism, was to base mathematics on an axiomatic treatment of logic. This was the method Frege and later Russell and Whitehead. Another more radical approach, known as Intuitionism, was taken by Brouwer. This called for a complete reworking of mathematics from a constructive viewpoint with the possible abandonment of fairly extensive portions of classical mathematics.

But the method most relevant to our considerations was that due to Hilbert. Hilbert wished to use metamathematical reasoning to guarantee the consistency of mathematics. He wanted to keep the theory of infinite sets and all of classical mathematics but also to avoid the paradoxes. His idea was that all of mathematics should be cast in the form of a formal axiomatic system. He believed that logic and mathematics went hand in hand so he intended to axiomatize them together.

Since the consistency of the various branches of mathematics could be ultimately reduced to that of arithmetic he thought that, as a first step, arithmetic should be axiomatized. The axiomatization was to be formal in that the axioms were to be devoid of meaning and the logical reasoning merely to consist of rules for producing one meaningless formula from another. He then proposed to prove the consistency of this logical structure by investigating all the possible proofs in the theory and thereby showing that it was impossible to prove two contradictory theorem, U and $\neg U$. The metamathematical reasoning used in this task had to be beyond reproach and for this he used what he called finitary reasoning. This secure form of reasoning was closely related to the intuitionistic reasoning which Brouwer wished to use for all of mathematics. We will consider Hilbert's metamathematics in more detail in Chapter 3.

1.4. Concluding Remarks

We have seen that the task of analyzing the concept of number led some mathematicians to adopt the notion of recursive definitions for arithmetic functions and led others to isolate the properties now referred to as the Peano axioms.

We have also seen how the rise of the axiomatic method associated with the desire to instil rigor into mathematics, and the change of status of the axioms led to consideration of the consistency of these axiom systems. The discovery of the paradoxes meant that proving this consistency was a vital concern.

Of the methods that were initially proposed for securing the foundations of mathematics the one most relevant to this book is Hilbert's metamathematics because it was to lead to a series of new results in recursive function theory. The detailed formulation of Hilbert's metamathematics did not occur, however, until the 1920's and we shall first consider Skolem's contributions to recursive function theory as presented in a paper which he wrote in 1919, but which was not published until 1923.

CHAPTER 2

Skolem's Contribution

2.1. Introduction

In this chapter we discuss Skolem's 1923 paper, "The Foundations of Elementary Arithmetic Established by Means of the Recursive Mode of Thought, Without the Use of Apparent Variables Ranging over Infinite Domains" [**153**] and illustrate its significance for the development of recursive function theory.

Skolem was the founder of recursive arithmetic. In this seminal paper he took the recursive method of defining functions, which, as discussed in the first chapter, had previously only been used to help understand the nature of numbers, and used it as the major part of his study of the foundations of arithmetic. The part of arithmetic that he developed has subsequently been called recursive arithmetic. Goodstein has appraised Skolem's work in the following terms:[1]

> The idea [of definition by recursion] first appeared explicitly in the work of Richard Dedekind, but the first to appreciate and exploit the full power of the method was Thoralf Skolem, one of the greatest architects of modern logic.

Unfortunately most of Skolem's work on logic was published in obscure journals and so was not known for a considerable period after publication. Consequently

---

[1]See [**61**, p. 59].

his 1923 paper had little influence on immediately succeeding logicians. In 1934 Hilbert and Bernays, in *Grundlagen der Mathematik* [**80**], gave a detailed description of Skolem' s work, their interest in Skolem' s arithmetic being due to the belief that it represented an adequate formulation of their finitary arithmetic. Later on Goodstein and Curry were also influenced by Skolem's work. Since Skolem's paper was the first on recursive arithmetic and since it contained some major developments in recursive function theory, it is worth considering in some detail.

Skolem was led to write this paper in the autumn of 1919 after reading Russell and Whitehead's *Principia Mathematical* [**174**]. He decided that a large part of arithmetic could be developed without their use of the notions "always" and "sometimes". Since existence was not of the primitive notions of his arithmetic Skolem was able to avoid the paradoxes and yet not get involved with the complexities associated with Russell and Whitehead's theory of types. His system of arithmetic was to be built up in a clear and constructive manner, the reasoning was to be so certain that existence of paradoxes within the system would be unthinkable. As he himself said in his conclusion:[2]

> ... [the work] is a consistently finitist one; it is built upon Kronecker's principle that a mathematical definition is a genuine definition if and only if it leads to the goal by means of a *finite* number of trials.

2.2. The 1923 Paper

In his paper Skolem defined the functions he needed for his arithmetic and proved many theorems using them. We shall now examine various parts of his paper in more detail.

---

[2]See [**64**, p. 333].

Skolem begins the introduction to his paper with a list of fundamental notions of logic which had been used in *Principia Mathematica* and were generally regarded as necessary for the foundations of mathematics:

   i) proposition and propositional functions of one, two or more variables,

   ii) the operations of conjunction, disjunction and negation

   iii) the qualifiers "always" and "sometimes"

   iv) descriptive functions

   v) functional assertions.

He then discusses his interpretation of some of these notions and his rejection of others.

He considers that all functions are descriptive functions, where he regards a descriptive function as a functional proper name whose meaning is dependent upon what numbers are chosen for the variables involved. For instance, the descriptive function $n + 1$, is the name of a number, though one that depends on the number that is chosen for n.

He means by a functional assertion a proposition that holds for the indeterminate case, for example:

$$a + b = b + a \text{ for any } a \text{ and } b$$

is a functional assertion. This functional assertion is equivalent, from the point of view of derivation and interpretation, to its universal closure

$$\forall a \ \forall b \ (a + b = b + a).$$

Skolem did not wish to use Russell and Whitehead's concepts of "always" and "sometimes". Rather, he wished to avoid the use of the unbounded universal and existential quantifiers altogether. Hence universality is only expressed by means of free variables as seen above and existence is not one of his primitive notions nor can an assertion of existence be the negation of a universal proposition. Consequently it is impossible to express the paradoxes in his arithmetic.

By his sacrifice of existence as a primitive notion Skolem deprives himself of the classical method of function definition, so in its place he introduces definition by recursion. He states that:[3]

> If we consider the general theorems of arithmetic to be functional assertions and take the recursive mode of thought as a basis, then that science can be founded in a rigorous way without use of Russell and Whitehead's notions "always" and "sometimes". This can also be expressed as follows: A logical foundation can be provided for arithmetic without the use of apparent logical variables.

By the recursive mode of thought he means two things:

   i) all definitions are definitions by recursion
  ii) all proofs are proofs by mathematical induction.

He finishes his introduction by giving his own assumed fundamental concepts:

    i) the natural numbers
   ii) the descriptive function, successor, $n + 1$
  iii) the substitution of equals for equals

to be taken along with the notions given previously. That is, when considering the general theorems of arithmetic as functional assertions and using the recursive mode of thought as a basis.

---

[3]See [**64**, p. 304].

During the rest of the paper he proceeds informally. He does not work within a particular formal system though he frequently uses the full power of sentential logic. He often uses words to express his logic and uses Schröder's notation throughout his paper.

Skolem does not give a general format for a recursive definition of a function. Such a definition was not forthcoming until Gödel's 1931 paper. He does, however, give plenty of particular examples such as:[4]

> I wish to introduce a description function of two variables, a and b which I shall call the *sum* of a and b. I denote it by $a + b$ since for $b = 1$ it is to mean precisely the number following a, namely $a + 1$. This function is therefore to be regarded as already defined for $b = 1$ for arbitrary a. To define it generally, I then need only define it for $b + 1$ for arbitrary a if it is already assumed to be defined for b for arbitrary a. This is accomplished by the following definition:
>
> $$a + (b + 1) = (a + b) + 1$$
>
> and
>
> $$a \cdot 1 = a \text{ and } a \cdot (b + 1) = a \cdot b + a.$$
>
> This is a recursive definition of a descriptive function $a \cdot b$ of two variables, a and b, which is called the *product* of a and b.

These two definitions are certainly not new, both Dedekind and Peano would have recognized them.[5]

---

[4]See [**64**, p. 305ff].

[5]As we have seen, Skolem partly specifies two initial functions in his introduction, namely, the successor function, $n + 1$, and the constant function, "the natural numbers," but he does not always show how his new functions are built up from these and previous definitions. Nor does he talk in terms of a class of recursive functions. It was not until Gödel's 1931 paper that these things were made precise.

The first new recursive definitions he gives are those of relations. Since Skolem saw his paper as a sequel to Russell and Whitehead's *Principia Mathematica* he always starts by stating their definitions in terms of existential or universal quantifiers and then shows how to avoid these quantifiers.

The first of his new definitions is that of the relation "less than". The definition in *Principia Mathematica* is given as:[6]

$$a < b = \exists x \, (a + x = b)$$

while his new recursive definition is

$$a < 1 \text{ is false and } a < b + 1 = (a < b) \lor (a = b)$$

Today the more usual method of defining recursive relations is in terms of their representing functions, but again this was not done until Gödel's 1931 paper.

Another example of the recursive definition of relations is his definition of the propositional function $D(a, b)$, "a is divisible by b".

He first defines

$$\Delta(a, b, 1) \overset{\text{def}}{=} (a = b)$$

$$\Delta(a, b, c + 1) \overset{\text{def}}{=} \Delta(a, b, c) \lor (a = b(c + 1))$$

so that

$$\Delta(a, b, c)$$

means that

"a is equal to b multiplied by a number between 1 and c".

---

[6]Skolem uses the equal sign in these formulæ in the form "is logically equivalent to" or "has the same truth value as". See [**64**, p.307]. From now on $\overset{\text{def}}{=}$ will be used.

Then
$$D(a,b) \stackrel{\text{def}}{=} \Delta(a,b,a).$$

Skolem makes the point that this definition not only avoids the quantifiers in Russell and Whitehead's definition of

$$D(a,b) \stackrel{\text{def}}{=} \exists x(a = bx)$$

but also avoids the definition

$$D(a,b) \stackrel{\text{def}}{=} (\exists x \le a)(a = bx)$$

using a bounded quantifier. Because of this bounded quantifier the apparent variable used in this second expression only ranges over a finite domain and, since this furnishes a finite criterion of divisibility, this is actually allowed in Skolem's arithmetic. At this stage in the paper Skolem always eliminates not only the quantifier involved but also the bounded quantifier which, in his arithmetic, could be used in its place. Later in the paper, for the sake of giving a shorter account of the theory, he chooses not to eliminate these bounded quantifiers. However, where he does eliminate the bounded quantifiers, he eliminates them as they occur, without indicating any generally applicable methods.

In Section 5 of the paper he defines a function with a restricted domain of existence. He produces the function $c - b$ by defining the relation $c - b = a$, that is:

$$(c - b = a) \stackrel{\text{def}}{=} (c = a + b)$$

where for $c \le b$, $c - b$ is left undefined.[7]

When he comes to define the functions "the greatest common divisor" and "the least common multiple" in Section 6 he finds he needs to extend his style of recursive definition. He defines $G(a,b)$ "the greatest common divisor of a and b", in terms of $G(a - b, b)$ and $G(a, b - a)$.[8] In other words, instead of always

---

[7]See [**64**, p. 314].

[8]He uses the notation $a \wedge b$ for the greatest common divisor of a and b.

defining a function for argument $n+1$ in terms of the function for argument n, he introduces course-of-values recursion as it is called today and definition by cases.

In the subsequent theorems he also uses the equivalent extended form of proof by mathematical induction. He indicates his conception of the possibility of going outside the class of simple recursive functions by showing, at the end of this section, that these new, more involved kinds of recursive definitions and inductive proofs differ only formally, and not actually, from the ordinary simple recursive procedures.

Having made a special point of the fact that these new definitions do not constitute a different sort of recursion, be then quite casually introduces a doubly recursive relation in the next section, Section 7. After defining the relation he explains that the definition utilizes double recursion and shows that it is constructive,, but he does not give any proof of equivalence as he had done previously for the course-of-values recursion. Whether this is because he considered the equivalence to be obvious is not clear. It may be that, even though the definition is recursive on two variables at once, the process still went from n to $n+1$ and so conformed with Skolem's principles concerning recursive definitions. Or, perhaps, he just accepted it as a different sort of recursion from the ordinary simple kind. In any case, he certainly shows that it is constructive so it is in the spirit of his paper. The definition in question is

$P(x, y, 1)$ is false

$$P(x, 1, z) \overset{\text{def}}{=} x \le z \ \& \ P(x)$$

$$P(x, y, z) = P(x, y, z - 1) \vee \left[P\left(\frac{x}{z}, y - 1, z\right) \ \& \ P(z) \ \& \ D(x, z) \right]$$

where $P(x)$ means "x is prime" and $D(x, z)$ means "x is divisible by z".

This definition can be converted into the more usual modern definition of a recursive representing function $p(x, y, z)$, so that $p(x, y, z) = 0$ if and only if

$P(x, y, z)$ is true, by first defining the Predecessor function $pd(x)$:

$$pd(0) = 0$$
$$pd(x + 1) = x$$

and the modified difference function:

$$x \mathbin{\dot{-}} 0 = x$$
$$x \mathbin{\dot{-}} (y + 1) = pd(x \mathbin{\dot{-}} y)$$

so that the modified difference function $x \mathbin{\dot{-}} y$ means ordinary difference if $x >= y$ and zero otherwise. Now $p(x, y, z)$ can be defined by:

$$p(x, y, 1) = 1$$
$$p(x, 1, z) = sg(x \mathbin{\dot{-}} z) + p(x)$$
$$p(x, y + 1, z + 1) = [p(x, y + 1, z)] \, [p(q(x, z), y, z + 1) + p(z) + r_m(x, z)]$$

where $sg(x)$ is the function defined by

$$sg(0) = 0$$
$$sg(x + 1) = 1,$$

$q(x, z)$ is the quotient function, $r_m(x, z)$ is the remainder function, and $p(x) = 0$ is the representing function for the relation "x is prime".

From this more modern definition of a recursive function it can be seen that, although the function is doubly recursive, the recursion is not nested and so it is still an ordinary recursive function.[9] The relation in Skolem's paper is therefore still an ordinary recursive relation and so Skolem had not stumbled onto a different kind of recursive definition, or a non-primitive recursive definition as it would be called today.

---

[9]This sort of analysis in terms of nested recursion was not completed until Rósza Péter's work in the 1930's.

Once Skolem has given this definition he proves various theorems on prime numbers up to the theorem that every number greater than one can be expressed as a product of primes.

In the following section, Section 8, he abandons his restriction on the use of bounded quantifiers and uses them to give a much simpler proof of this last theorem, and then continues by proving that this factorization is unique.

During this work he defines another doubly recursive but still ordinary recursive relation and also defines functionals. For example:

$$\sum_{r=1}^{1} f(r) = f(1)$$

$$\sum_{r=1}^{n+1} f(r) = \sum_{r=1}^{n} f(r) + f(n+1)$$

$$\prod_{r=1}^{1} f(r) = f(1)$$

$$\prod_{r=1}^{n+1} f(r) = \prod_{r=1}^{n} f(r) \cdot f(n+1)$$

Finally, for an arbitrary propositional function U. He defines the descriptive function $\mathrm{Min}(U, n)$. This function $\mathrm{Min}(U, n)$ is defined to be the least number among the numbers 1 to n for which U is true. If U is false for all these numbers then $\mathrm{Min}(U, n)$ has no meaning. He then carefully explains that if he had defined the function $\mathrm{Min}(U)$ instead, and U was false for every number, then he would be involved in the "actual infinite". Hence, to conform with his finitist intentions, he defines the function $\mathrm{Min}(U, n)$. He claims that in practice the function $\mathrm{Min}(U, n)$ is quite sufficient since he wishes to use the minimum function only in the circumstances where he has already established a value of n for which U is true.

2.3. Comments and Conclusion

From this study of Skolem's paper we can see that Skolem extended the use of recursive definitions considerably. He gave the first recursive definitions of propositional functions (relations) and he introduced recursion as a method of eliminating bounded quantifiers. He also introduced course-of-values recursion and definition by cases and finally he used double recursion. All these developments indicate a substantial improvement in the use made of recursive functions from that of Dedekind and Peano.

Skolem based his arithmetic on recursive functions. For him they were vital to his exposition in that they allowed him to develop his arithmetic constructively. Dedekind and Peano, on the other hand, used recursive definitions as just a part of their theories. Dedekind based his arithmetic on set-theoretic notions and justified his recursive definitions in these terms and Peano used recursive definitions as a part of the process of producing an axiomatic foundation for arithmetic.

Throughout the paper Skolem proceeded informally, introducing functions and methods as he needed them, and consequently he did not give any general methods. He did not specify a set of initial functions nor a set of rules by which he could generate all his recursive functions.

He did not think in terms of a class of recursive functions at all but always argued each case individually. It was not until 1931 that Gödel gave the first precise definition of the class of recursive functions, but this precise definition incorporated the assumed notions used by Skolem, namely:

i) the natural numbers
ii) the successor function, $n + 1$
iii) substitution of equals for equals
iv) the recursive mode of thought.

Skolem did not give any general methods of eliminating bounded quantifiers: a few specific examples were given and then in Section 8, he argued that he could use bounded quantifiers freely since it could be seen directly that they were acceptable. For these general methods we had to wait for Gödel.

Skolem's limited examination of other, more complex, methods of defining recursive functions and the reductions of these to ordinary recursive form could not be discussed constructively until a more precise definition of the class of recursive functions was forthcoming. Ackermann, Gödel and Péter were all to make considerable contributions to this task, but nevertheless it was Skolem who was the first to deal with such alternative definitions.

Skolem thought that his system of arithmetic ought to settle all the arguments on the foundations of arithmetic, as can be seen from this excerpt from an address he gave to a conference in Helsinki in 1922.[10]

> It is odd to see that, since the attempt to find a foundation for arithmetic in set theory has not been very successful because of the logical difficulties inherent in the latter, attempts, and indeed very contrived ones, are now being made to find a different foundation for it – as if arithmetic had not already an adequate foundation in inductive inferences and recursive definitions.

This passage also, incidentally, shows Skolem's opinion about the unsatisfactory nature of the foundations of arithmetic afforded by set theory. His opinions, we stated more fully in the following extract from the same conference:[11]

---

[10]The conference was held before the actual publication date of his paper, but long after he had written it. See [**64**, p. 300].

[11]See [**64**, p. 300–301].

> I believed that it was so clear that axiomatization in terms of
> sets was not a satisfactory ultimate foundation of mathematics
> that mathematicians would, for the most part, not be very
> much concerned with it.

Skolem showed no knowledge of any of Hilbert's thoughts on the foundations of arithmetic or on metamathematics and did not consider any metamathematical questions, though the treatment of any such questions would necessarily have had to be simplistic due to the informal approach that he took. It is true that Skolem was in Göttingen in the years 1915–1916 but he was engaged in some work on set theory at the time.[12] In any case, Hilbert did not properly formulate his ideas on the foundations of arithmetic until after 1917. Before 1917 he was still mainly investigating the possible axiomatization of Physics. Skolem later admitted that he was also unaware of Brouwer's writings at the time[13] so that the similarity to intuitionism was purely coincidental. Skolem was, however, significantly influenced by Kronecker, who believed in the finite constructivity of all of mathematics and after Kronecker several other mathematicians including Brouwer took similar philosophical standpoints, so it is perhaps not really surprising that the two systems bear some resemblance to each other. Skolem's system is, in fact, more restrictive than intuitionism.

Skolem was always very careful only to use finite methods and, whenever a situation occurred in which a possible infinity could be produced, he explained clearly why his definition was acceptable. This has already been seen in his definition of $\mathrm{Min}(U, n)$ rather than $\mathrm{Min}(U)$. Skolem's 1923 paper, like so many others of his in logic, remained largely unnoticed and unread and so appears to have bad little influence on contemporary logicians. We will investigate any possible influence on Hilbert and Gödel in the introduction to the next chapter.

---

[12]Hilbert was Professor at Göttingen from 1895 to 1930.

[13]This statement was made at a mathematics congress in Oslo in 1929.

Some time later Goodstein and Curry extended and improved Skolem's ideas and produced more complete system of recursive arithmetic. Skolem's recursive theory has developed considerably since its inception but his very restricted construction of arithmetic has few adherents today.

CHAPTER 3

Hilbert's Program and Gödel's Incompleteness Theorems

3.1. Introduction

Hilbert's metamathematics was, as mentioned in the first chapter, one of the principal philosophies adopted to combat the paradoxes which arose from set theory. In this chapter we shall analyze Hilbert's approach to the consistency of arithmetic by considering his papers and the papers of those mathematicians who supported and followed him in his work. We shall trace the gradual development of a mature methodology up to the point where success seemed imminent, only to be dealt a death blow by Gödel's 1931 paper, although Hilbert himself continued to retain hope for his plan. From Hilbert's work and Gödel's paper were to come many new mathematical ideas, tools and procedures, not least of which was a rich fund of new results on recursive functions. It is because of these new developments in recursive functions and because Gödel's theorems, which had such a profound effect on foundational research, developed out of Hilbert's approach, that we shall consider Hilbert's formalist program in such detail.

Hilbert's and Gödel's use of recursive functions, raises the question of whether or not they knew of Skolem's work on recursive functions. As we shall see, both Hilbert and Gödel produced many results on recursive functions which were similar to Skolem's, but they also produced many that were different. In places it appears that Skolem's work was totally unknown while in others it appears to have been taken for granted. There are considerable differences of opinion as to whether Skolem's work was known of or used by Gödel or Hilbert or any of his co-workers. It was certainly known of by 1934, since Hilbert and Bernays give considerable reference to Skolem in Volume I of their book, *Grundlagen der Mathematik*. Hao Wang says in a letter of 16 July 1979 that he doubts if Hilbert or Gödel knew of Skolem's work. Kleene is of the same opinion as regards Gödel. As Kleene says in his letter to the author of 17 July 1979:

> It seems to me just unlikely that someone with as much momentum on his own research as Gödel would have spent very much time digging around in obscure places in the literature.

However, van Heijenoort believes that both Hilbert and Gödel did know of Skolem's work. He believes that Skolem's paper presented a point of view which was readily accepted by everyone but which did not contain any real deep technical results and was therefore just assimilated into the background knowledge of the subject, without ever meriting a credit in further papers. If this is so then it is perhaps not surprising that other researchers used similar ideas to Skolem's but chose slightly different definitions and approaches to suit their own purposes. Van Heijenoort backs his opinion by saying that Gödel made a point of informing him, when they met in Princeton, that he did not know of Skolem's 1923 paper prior to writing his own 1930 paper. Van Heijenoort infers from this that Gödel did know of the 1923 paper.

There is evidence in Hilbert's 1922 address at Copenhagen that, in fact, Hilbert was producing similar work before Skolem's paper was published and certainly Hilbert's work could all have been derived from the foundations laid down by Dedekind and Peano. It seems to me that Hilbert or his collaborators produced their own work on recursive functions independently at first and while, later on, they may have known of Skolem's work, they did not feel that it added enough to their own approach to credit him with the results.

Gödel, on the other hand, knew almost all of Hilbert's work and could have thought that the recursive function-theoretic results were Hilbert's results. He would see no need to credit Hilbert with this specifically since he acknowledges Hilbert's overall influence on his work. He would also not feel it was necessary to mention to van Heijenoort whether or not he did know of Skolem's 1923 paper since it was insignificant compared with the wealth of information he gained from Hilbert's school. That is, Gödel may have known of Skolem's work through Hilbert's papers but probably had not actually read the paper himself.

In this chapter we shall first consider Hilbert's motivation for his program and then discuss his methodology and general approach before considering his formalist program in some detail through an analysis of his papers.

3.2. Hilbert's Approach

One of Hilbert's many interests in mathematics was to make secure the foundations of all the branches of mathematics. He believed that the axiomatic method that was currently being practised was superior to the genetic method and he stressed its importance when giving a complete description of any branch of mathematics. In the 1890's Hilbert worked on the foundations of geometry and in 1899 he produced a classic axiomatic system for geometry in his book, *Grundlagen der Geometrie*. In 1900 he produced a set of axioms for the whole real number system. He preferred the axiomatic method for developing the real numbers because of its logical precision and the fact that it gave a definitive formulation.

Having axiomatized a theory it is then possible to examine various general questions such as consistency. Since truth was no longer considered to be an adequate justification for the correctness of a theory this was seen to be a rather important task. For the set of geometric axioms given in his 1899 book Hilbert in fact investigated several questions, such as the consistency and the independence of the axioms.

He solved the consistency problem by producing a model involving the set of real numbers using Cartesian co-ordinates and thereby showed that any contradiction that could occur in deductions from his geometric axioms must be derivable in the arithmetic of real numbers. This was a relative consistency result and several were known at the time. But these consistency proofs did not solve the problem of the consistency of the real numbers and when, shortly afterwards, the paradoxes were discovered, the problem of proving arithmetic, and therefore analysis, consistent became more urgent.

As mentioned in Chapter 1, in 1900 Hilbert placed the problem of consistency of arithmetic as the second of his twenty-three problems. He recognized that the method of proof must be different from the previous relative consistency proofs because there was no other suitable axiom system to fall back on. In 1900 he stated:[1]

> I am convinced that it must be possible to find a direct proof for the compatibility [consistency] of the arithmetical axioms, by means of a careful study and suitable modification of the known methods of reasoning in the theory of irrational numbers.

He continued:

> But if it can be proved that the attributes assigned to the concept can never lead to a contradiction by the application of a finite number of logical processes, I say that the mathematical

---

[1]See [**69**, p. 448].

existence of the concept (for example, of a number or a function which satisfies certain conditions) is thereby proved. In the case before us, where we are concerned with the axioms of real numbers in arithmetic, the proof of the compatibility of the axioms is at the same time the proof of the mathematical existence of the complete system of real numbers or of the continuum. Indeed, when the proof for the compatibility of the axioms shall be fully accomplished, the doubts which have been expressed occasionally as to the existence of the complete system of real numbers will become totally groundless."

This shows that Hilbert believed the task of proving the consistency of arithmetic to be extremely important.

Hilbert gradually unfolded a new method for dealing with this problem. In his 1899 treatise on geometry he produced an abstract axiom system in that the intended meaning of the geometric terms contained in the theory could not enter into the derivations from that theory. His was not the first such formal axiom system for geometry. Pasch in 1882 [115] was probably the first. But it was one of the first and Hilbert's reputation meant that it was almost certainly the most influential. The meaning of the geometric terms had become irrelevant but the meaning of the logical terms had still to be understood.

When Hilbert turned his attention to arithmetic he made two major contributions to the solution of the consistency problem:

i) the idea of totally abstracting from the meaning of, not only the primitive terms of the theory but also the logic used, giving a completely formal structure, called a formalism, where deductions could only be made on the basis of the formal axioms and formal rules on the shape of the logical sentences,

ii) the taking of the formal system as a whole as the object of a mathematical study, the purpose of this study being to show that application
of the rules of inference to the axioms could never lead to a formal contradiction. By this he meant that there would be no way of producing
a derivation of a sentence U and also a sentence $\neg U$.

The idea described in (ii) above, where the proofs of the formal theory became the objects in some informal theory, is called metamathematics or proof
theory. Obviously the type of argument allowed in this metamathematics must
be beyond reproach and so Hilbert specified that only finitary reasoning should
be used. Hilbert, unfortunately, never completely specified what procedures
were to be regarded as finitary, but the arguments were basically more strict
than even those arguments used by the Intuitionists. These ideas were rather
vaguely presented in his 1904 paper but became more precise after 1922 when
Hilbert returned with a vengeance to the problem of the consistency of arithmetic for a more personal reason: that of defending mathematics against the
"Ghost of Kronecker," the Dutch mathematician, Brouwer.

In his early years Hilbert had championed the cause against Kronecker's
views, especially against the belief that existence theorems should be proved by
explicit construction in terms of integers. Hilbert saw Brouwer's Intuitionism
as Kronecker's beliefs being resurrected but this time with much more force
and with an increasing degree of support. Someone was needed to fight against
Brouwer to stop him destroying large sections of the mathematics that Hilbert
knew and loved. This task Hilbert took upon himself and from 1922 onwards
he formulated his defence of arithmetic.

Utilizing his early ideas of 1904 he expanded, clarified and improved them
to such an extent that he succeeded in presenting a viable alternative view to
that of Brouwer. Mathematicians saw hope in Hilbert's plan and, aided by a
younger group of collaborators whom he inspired and led on with great conviction, he started proving the consistency of axiom systems of small fragments

of arithmetic, gradually improving and enlarging the system in the hope of eventually encompassing all of arithmetic and analysis.

Unfortunately that ultimate objective was never realized. Gödel's 1931 paper showed that Hilbert's plan, as he had originally envisaged it, could never be completed. But Hilbert's defence of mathematics has led to many new results and methods which have transcended their original purpose and it also greatly reduced the appeal and support for Intuitionism.

We can see some of these ideas developing by considering Hilbert's papers and addresses in more detail. At the same time we will observe an increasing reference to recursive functions as Hilbert expands the role for these functions.

3.3. Hilbert's Foundational Papers

Hilbert's 1904 paper is the text of the address he delivered on 12 August at the Third International Congress of Mathematicians at Heidelberg, and contains the first of his attempts to prove the consistency of arithmetic. In this paper we see the first statement of his intention to develop the axiomatization of arithmetic and logic simultaneously. This approach of Hilbert's is contrary to the views of philosophers like Frege, who believed in basing mathematics on logic, as is his belief in the extralogical existence of thought objects such as the number 1 and their combinations. He needs to invoke a sort of global intuition to allow for recognition of differences in combinations of these thought objects.

Having stated his assumptions, Hilbert embarks on the first direct proof of the consistency of a small set of axioms. He achieves this by showing that the axiom and rules of inference (these rules are rather inadequately defined) do not lead to a contradiction, that is he proves that both the proposition P and the proposition $\neg P$ can not, at the same time, be consequences from the axioms. His method of proof, which has been sharpened and used frequently in subsequent investigations of this nature, consists of arguments based on the "shape" of the sentences involved and consequently is suitable for uninterpreted meaningless fomulæ.

Unfortunately during this proof Hilbert had, unknowingly used mathematical induction, that is he used the argument that if, after $n + 1$ uses of the axioms and rules of inference he had not got a contradiction, assuming that he had not already got a contradiction after n uses, then he would never get a contradiction. Poincaré pointed this out in a severe criticism of the paper in an article [124]–[125] in the *Revue de métaphysique et de morale* of 1905. Hilbert was to reply to this criticism in a later paper by dividing his arguments into two levels, formal and metamathematical, so that he could use two forms of induction, though he was far from making this clear in 1904.

Hilbert thought at the end of this paper that he had indicated a method of proving absolute consistency of a set of axioms and he obviously hoped that someone would take up the challenge and complete the work while he himself went on to follow other interests. Unfortunately, because the paper was rather vague and tentative and because it generated rather a lot of criticism, especially from Poincaré, hardly anyone took up the challenge.[2]

Hilbert's next paper on this topic is the text of an address given on 11 September 1917 in Zürich and is a general discussion paper. He restates his interests in the foundations problem and discusses the various solutions to the troubles caused by the paradoxes. He indicates that an axiomatic approach seems to be the way forward but he does little more than show that he is watching developments.

---

[2]Julius König did attempt to carry out Hilbert's plan but he died before he could complete very much. Hilbert did not know of König's work when he returned, in 1922, to investigate the consistency of arithmetic. König's work, edited by his son, can be found in [99].

His next two papers are addresses given in 1922 and were written in the spring and the summer of that year. They were written to defend mathematics against the attacks by Brouwer and were ultimately precipitated by his friend Weyl having recently joined Brouwer's camp and having delivered speeches in support of intuitionism. These two papers represent the start of his serious attempts to solve the foundations problem and contain the salient features of his plan to save mathematics. The first paper contains some axioms for arithmetic and logic and mentions his method of proving consistency, namely by showing that both $a = b$ and $a \neq b$ can not be proved in his formal system. The paper contains Hilbert's reply to Poincaré's criticism, in which he suggests setting up two sorts of induction, and also contains the first mention of metamathematics.

By the second paper of 1922 Hilbert's ideas are better formulated; he again gives a set of axioms for his formal system and he also indicates that the definitions of functions via recursion should be incorporated into this formal system. He discusses the principles underlying his metamathematics and suggests the use of recursion and intuitive induction as part of the finitary reasoning used in his metamathematics, though by now his analysis of the proofs in the formal system are supposed to show that the end formula $0 \neq 0$ can not be produced.

This indication of the use of recursion in his finitary reasoning occurs before Skolem's 1923 paper was published, as did the first use of simultaneous recursion to define functions, namely:

$$\chi(0, n) = 0$$
$$\psi(a + 1) = \pi_n \{ \chi(a, n) = 0 \implies \phi(a + 1, n) = 0 \}$$
$$\chi(a + 1, n) = \chi(a, n) + \iota(\psi(a + 1), \phi(a + 1, n))$$

where ψ and χ are the two functions that are being defined simultaneously in terms of the functions π_n, $\iota(a, b)$ and $\phi(a + 1, n)$ which had previously been defined. Simultaneous recursion had not been used by anyone before Hilbert and, in fact, was not even used by Skolem. The existence of this definition in Hilbert's paper is part of the evidence that Hilbert was producing his own recursive function theory results rather than using Skolem's work.

Hilbert's next two papers are also texts of addresses. The 1925 paper is an address given on 4 June 1925 at Münster and the 1927 paper is an address given in July 1927 at Hamburg. They both contain very clear descriptions of Hilbert's ideas and hopes on the foundations of arithmetic and start with his admission that mathematics had gone beyond ordinary contentual thinking and that even elementary mathematics had gone beyond this by using and manipulating algebraic formulæ, since these formulae represent infinite propositions.

He states that:[3]

> It is necessary to make inferences everywhere as reliable as they are in ordinary elementary number theory, which no one questions and in which contradictions and paradoxes arise only through our carelessness.

He argues that to abandon the powerful methods of argument utilizing non-finitary algebra would be madness and that we can avoid doing this by considering all those non-finitary propositions as ideal elements in a theory so that the simple rules of ordinary Aristotelian logic can be maintained. But this step would require the proof of the consistency of the enlarged system containing the ideal elements, that is, it would be necessary to show that this extension into ideal elements did not bring any inconsistencies into the old narrow domain.

This proof of consistency could be carried out by the techniques he had already indicated in his previous papers, but which he explains again in more detail in these papers. This method of proof would also give the long-desired consistency of arithmetic relative, of course, to the finitary arithmetic used in the metamathematical argument.

---

[3]See [**64**, p. 376].

He had complete confidence in the ultimate success of his plan to prove the whole of arithmetic consistent as can be seen in the following quotation:[4]

> ... it is possible to obtain in a purely intuitive and finitary way, just like the truths of number theory, those insights that guarantee the reliability of the mathematical apparatus.

Also included in these last two papers is a brief attempt by Hilbert at indicating a proof of the continuum hypothesis. This attempt is rather fragmentary and contains several features that would be a hindrance to its successful completion. It was, however, responsible for setting in motion a thorough investigation of recursive functions and an indication of the reasons for this will be given in the next section.

3.4. Results on Recursive Function Theory

These last two papers both contain a number of stated results on recursive functions that were proved by Hilbert and his collaborators during the 1920's. These recursive function theory results were produced to further both the proof of the consistency of mathematics and the proof of the continuum hypothesis.

With regards to the consistency of mathematics, Hilbert proposed to make a metamathematical study of all the possible proofs that could produced in a formal structure intended to represent arithmetic. The reasoning used was to be immediately intuitive and directly intelligible. It was soon realized that recursive methods represented a possible framework for this "finitary" arithmetic. This conclusion can clearly be seen from the following passage taken from Hilbert's 1925 paper, where he is considering whether recursion and substitution would be all that was needed for his definition of functions.[5]

---

[4]See [**64**, p. 377].
[5]See [**64**, p. 388].

> The method of search for the recursions required is in essence
> equivalent to that reflection by which one recognizes that the
> procedure used for the given definition is finitary.

The continuum hypothesis, first stated by Cantor in 1878, basically claims that the totality of all real numbers can be enumerated by using the ordinals of the second number class. Hilbert recognized that this was an important result and included it as his first problem in his list of twenty-three problems in 1900. He attempted to prove it in his 1925 and 1927 papers.

He first considered the set of all number-theoretic functions instead of the equivalent set of all real numbers. He then needed to associate an ordinal of the second number class with each number-theoretic function. Hilbert decided to deal with the possible methods of definition of all the functions rather than with the functions themselves. A study of the foundations and possible methods of generation of all functions led naturally to recursive functions, since these functions could all be generated in order from very simple beginnings.

Once Hilbert had concluded that substitution and recursion were the only methods of definition that he would need he could then proceed to use recursions of more and more complicated kinds and associate higher and higher transfinite numbers with the functions that he thus obtained. He introduced Ackermann's function (defined in section 3.4.4) to prove that more complicated kinds of recursion did give a wider class of functions and would therefore be needed in his proof. His use of different kinds of recursion left open the question as to precisely which kinds of recursion define the same class of functions and which define different classes of functions. The theory associated with Ackermann's function represented a start to this investigation, but, as we shall see, the main outcomes were to be Péter's classification of recursive schemas and the introduction of general recursive functions by Gödel and Kleene.

Hilbert's program, for proving the continuum hypothesis has never been carried through, though Gödel did use an analogous method, omitting some of Hilbert's restrictions, in 1938 when he proved that the continuum hypothesis was consistent with the usual axioms for set theory. Hilbert's attempt was not very well understood at the time[6] and Freudenthal, in the *Dictionary of Scientific Biography* [**47**, p. 392–393] describes Hilbert's work in this field as poor and shallow compared with his other mathematical work. Yet Hilbert's use of different kinds of recursion was to have far-reaching effects.

We can therefore see that recursive functions played a central role in Hilbert's papers of the 1920's and in various papers by his team of helpers. Some of the actual advances made during this period can be categorized as follows:

  i) recognition and a clarification of the overall scope of the functions
  ii) a definition of transfinite recursion
  iii) a production of a general schema for ordinary recursion
  iv) an extension of ordinary recursion.

3.4.1. Scope of the Functions. As we have seen, Hilbert had already recognized that recursive definitions and substitutions were the correct formal framework for metamathematics. He claimed in various of his papers that it was well known that the functions sum, product, factorial, remainder in division, greatest common divisor of two numbers and the least value of finitely many numbers could be defined by recursion and substitution alone, as could the concepts of prime number and the number of primes below a certain number. In his paper of 1925 Hilbert considers whether other elementary methods of definition are needed to define all number theoretic functions, but concludes that recursion and substitution alone will suffice to generate all functions.

---

[6]See [**101**, p. 89] for comments along this line.

3.4.2. Transfinite Recursion. Hilbert defines transfinite recursion as follows:[7]

> The generalization of ordinary recursion is transfinite recursion; it rests upon the general principle that the value of the function for a value of the variable is determined by the preceding values of the function.

This allowed him to define recursion for functions whose arguments were numbers of the second number class.

Transfinite recursion figures in two controversial lemmas given in Hilbert's 1925 paper, namely:

> *Lemma 1.* If a proof of a proposition contradicting the continuum theorem is given in a formalized version with the aid of functions defined by means of the transfinite symbol ϵ, then in this proof these functions can always be replaced by functions defined, without the use of the symbol ϵ, by means merely of ordinary and transfinite recursion, so that the transfinite appears only in the guise of the universal quantifier.[8]

> *Lemma 2.* In the formation of functions of a number-theoretic variable transfinite recursions are dispensable.[9]

---

[7]See [**64**, p. 386].

[8]Hilbert called ϵ the transfinite logical choice function and it was defined by the axiom: $A(a) \rightarrow A(\epsilon(A))$, where A is a propositional variable. So that $\epsilon(A)$ stood for an object of which the proposition $A(a)$ certainly held if it held for any object at all. It can be shown that this axiom is the only one that is needed to produce all the transfinite axioms containing universal and existential quantifiers. See [**64**, p. 385].

[9]See [**64**, p. 391].

Neither lemma is proved.

The first lemma shows Hilbert's conviction that he could justify the trans-finite part of mathematics by finitary metamathematical reasoning. In his next paper Hilbert claims that Lemma 1 is dispensable and adds some explanations to Lemma 2, but it still remained a hopeful statement, far from being proved.

3.4.3. General Schema for Ordinary Recursion.
In his paper of 1925 Hilbert claims that certain functions depart from ordinary stepwise recursion, for example:

$$\phi_0(a) = \mathfrak{a}(a)$$
$$\phi_{n+1}(a) = \mathfrak{f}(a, n, \phi_n(\phi_n(n+a)))$$

or equivalently

$$\phi(a, 0) = \mathfrak{a}(a)$$
$$\phi(a, n+1) = \mathfrak{f}(a, n, \phi(\phi(n+a, n), n)).$$

The indications are that he believed this type of function led out of the class of ordinary stepwise recursions and so be gave a general schema for recursions on ordinary primitive variables that would include the above function and other functions of a similar nature. This general schema is as follows:[10]

$$f(x_1, x_2, \ldots, x_k, 0) = g(x_1, x_2, \ldots, x_k)$$
$$f(x_1, x_2, \ldots, x_k, n+1) = \mathfrak{h}_{y_1, \ldots, y_k}(x_1, \ldots, x_k, n, f(y_1, \ldots, y_k, n))$$

where $\mathfrak{h}$ is a function obtained by composition from the initial functions, previously obtained ordinary recursive functions, and the function $f(y_1, \ldots, y_k, n)$ is considered as a function of $y_1, y_2, \ldots, y_k$. The subscripts $y_1, \ldots, y_k$ merely mark argument places of f but do not actually occur in the function f; the

---

[10]See [2, p. 503] and [119, p. 614]. The page reference in Ackermann's paper refers to the translation in [64].

argument places marked by $y_1, \ldots, y_k$ have been occupied by functions of the kind just prescribed.

To see that the function given at the beginning of this section can be expressed using Hilbert's general schema we can write $y_1 = \phi(n+a, n)$ and obtain

$$\phi(a, 0) = \mathfrak{a}(a)$$
$$\phi(a, n+1) = \mathfrak{h}_{y_1}(a, n, \phi(y_1, n))$$

which is Hilbert's ordinary recursive schema with one parameter a.

We can also illustrate Hilbert's general schema by considering another example, namely:[11]

$$\psi(a, 0) = a + 1$$
$$\psi(a, b+1) = 1 + \psi(\psi(a+b, b), b)$$

writing

$$f(a, b, \psi(\psi(a+b, b), b)) = 1 + \psi(\psi(a+b, b))$$

gives

$$\psi(a, 0) = a + 1$$
$$\psi(a, b+1) = f(a, b, \psi(\psi(a+b, b), b))$$

and writing

$$g(a) = a + 1 \quad \text{and} \quad y_1 = \psi(a+b, b)$$

we obtain

$$\psi(a, 0) = g(a)$$
$$\psi(a, b+1) = \mathfrak{h}_{y_1}(a, b, \psi(y_1, b))$$

which is Hilbert's general schema for ordinary recursive functions.

—————————

[11]See [**2**, p. 503].

This general schema of Hilbert's appeared to give a much wider class of functions than the ordinary stepwise recursions of Skolem, but in 1934 Péter proved that Hilbert's schema and Gödel's schema, which corresponded to Skolem's original format but was given in precise form, defined exactly the same set of functions which we now call primitive recursive functions - so the extra freedom gained by using variables $y_1, y_2, \ldots, y_k$ was purely illusory.

3.4.4. Extension of Ordinary Recursion. Hilbert defined and used variable types to extend his ordinary recursive schema.

Hilbert, in his 1925 paper, cites Ackermann's function as proof that a real extension on ordinary stepwise recursion can be obtained.[12] Ackermann actually proved this result in his 1928 paper by showing that his new function increased more rapidly than all ordinary recursive functions and so must lie outside the class of ordinary recursive functions.

Obviously only one function was needed to show that it is possible to go outside the class of ordinary recursive functions, but both Hilbert and Ackermann showed how to generate a whole class of functions, each of which would lie outside this class. To illustrate this extended class of functions we will give Hilbert's general schema, since Ackermann did not provide one, but we will use Ackermann's notation, since it is slightly more general.

The construction of these functions uses the notion of variable types. A function of type one is one whose arguments and values are integers, that is

---

[12]Cristian Calude and Solomon Marcus [13] have argued the case that Gabriel Sudan ought to share with Ackermann the authorship of the first non-primitive recursive function. Sudan was with Ackermann and Hilbert in Germany from 1922 until 1925 when Sudan returned to Romania. Ackermann refers to Sudan's forthcoming paper in a footnote to his 1928 [2] paper where he shows that he is acquainted with the results contained in Sudan's paper. There is no doubt that Ackermann and Sudan knew of each other's work and published some similar results but there is no evidence to imply that it was not Ackermann who first thought of the idea.

ordinary number-theoretic functions. A function of type two is a function of a function, in that one or more of its arguments is a function of type one and not just a function value, its value being an integer. A function of type three has one or more of its arguments as a function of type two. This hierarchy can be continued into the transfinite and is part of Hilbert's process of numbering all functions.

An example of a function of type two which is recursively defined is the following:

$$\rho_c(f(c), a, 0) = a$$
$$\rho_c(f(c), a, n+1) = f(\rho_c(f(c), a, n))$$

This function is called the iteration function, an appropriate name, as can be seen from:

$$\rho_c(f(c), a, 0) = a$$
$$\rho_c(f(c), a, 1) = f(\rho_c(f(c), a, 0)) = f(a)$$
$$\rho_c(f(c), a, 2) = f(\rho_c(f(c), a, 1)) = f(f(a))$$

and so forth.

Having defined function types, all functions can then be classified according to whether or not their recursive definitions make essential use of functions of some given type and no higher types. For example, a function that utilizes a type two function in its definitions is Ackermann's function

$$\phi(a, b, 0) = a + b$$
$$\phi(a, b, n+1) = \rho_c(\phi(a, c, b), \alpha(a, n), b)$$

which uses the iteration function ρ_c.

In this context, $\alpha(a, n)$ is an ordinary recursive function defined in terms of two of Ackermann's initial functions, $\iota(a, b)$ and $\lambda(a, b)$ so that

$$\alpha(a, 0) = 0$$
$$\alpha(a, 1) = 1$$
$$\alpha(a, n) = a \text{ for } n \geq 2.$$

Ackermann, in his 1928 paper, shows that, if recursion is limited to one variable only, his function cannot be recursively defined without recourse to type two functions. However during the course of this paper Ackermann shows that his function can be alternatively defined by:

$$\phi(a, b, 0) = a + b$$
$$\phi(a, 0, n + 1) = \alpha(a, n)$$
$$\phi(a, b + 1, n + 1) = \phi(a, \phi(a, b, n + 1), n)$$

which uses simultaneous nested double recursion. This alternative form is acknowledged by Hilbert in 1925 as is shown by:[13]

> To be sure, we could now define $\phi_n(a, b)$ [that is $\phi(a, b, n)$] for variable n by means of substitutions and recursions, but these recursions would not be ordinary, stepwise ones; rather, we would be led to a manifold simultaneous recursion, that is, a recursion on different variables at once, and a resolution of it into ordinary, stepwise recursions would be possibly only if we make use of the notion of function variable.

Hence simultaneous double recursion also leads out of the class of ordinary recursive functions. To ensure the use of only singly recursive functions a recourse to higher type functions is needed.

---

[13]See [**64**, p. 388].

Hilbert even gives a general schema for recursively defining all functions in terms of a single recursive variable including those functions requiring higher type functions in their definitions, namely:[14]

$$\rho(\mathfrak{g}, \mathfrak{a}, 0) = \mathfrak{a}$$
$$\rho(\mathfrak{g}, \mathfrak{a}, n+1) = \mathfrak{g}(\rho(\mathfrak{g}, \mathfrak{a}, n), n)$$

> Here $\mathfrak{a}$ is a given expression of arbitrary variable-type; $\mathfrak{g}$ likewise is a given expression, having two arguments, of which the first is of the same variable-type as $\mathfrak{a}$ and the second is a number; the additional condition that $\mathfrak{g}$ must satisfy is that its value again be of the same variable-type as $\mathfrak{a}$. Finally, ρ is the expression to be defined by the recursion; it depends on three arguments and, after the substitutions for $\mathfrak{g}$, $\mathfrak{a}$ and n have been made, is of the same variable-type as $\mathfrak{a}$; in addition, other arbitrary parameters are permitted to occur in $\mathfrak{a}$ and $\mathfrak{g}$, and consequently also in ρ.

This general schema includes all of Hilbert's ordinary recursive functions, Ackermann's function and others defined similarly in terms of higher type functions.

This general schema does not appear in later literature but the discovery of Ackermann's function, and other functions like it, was to have far reaching effects: as has already been stated, it led to the classification of recursive schemas, this work being mainly undertaken by Rósza Péter in the 1930's and it was also to lead to the notion of general recursive functions. Péter's work will be addressed in detail in Chapter 5.

Finally, the issue of initial functions, which we have postponed discussing until now, needs to be clarified. In all the papers of the 1920's utilizing recursive

---

[14]See [**64**, p. 389].

functions the initial functions were left rather vague. It was generally recognized that certain start functions were required and that the list of such functions could be quite drastically reduced but no consensus was reached as to which initial functions should be used, the successor function being the only function that all authors recognized. Ackermann specifies the three initial functions:

i) the successor function, $a + 1$

ii) the function $\iota(a, b)$ defined by $\iota(a, b) = \begin{cases} 0 & a = b \\ 1 & a \neq b \end{cases}$

iii) the function $\lambda(a, b)$ defined by $\lambda(a, b) = \begin{cases} 1 & a = b \\ 0 & a \neq b \end{cases}$

He does admit, though, that λ and ι can be defined recursively, presumably in terms of more basic initial functions. By the 1930's, however, the position had changed since Gödel, in 1931 and 1934, was quite definite about which functions he would use and, following him, Péter, Kleene and R. Robinson have clarified the situation considerably.

3.5. The Fate of Hilbert's Plan

During the 1920's several important steps were taken towards achieving Hilbert's objective and there was considerable optimism that the hoped for consistency proof for analysis would soon be forthcoming. Ackermann, in 1924, undertook the task of converting Hilbert's indications of how to produce consistency proofs into a full consistency proof for analysis. He actually thought that he had succeeded but shortly before publication he realized that his proof was not adequate for analysis so he had to restrict his substitution rule to allow the proof to stand. The resulting formal system was then well short of analysis. In 1927 von Neumann discovered a consistency proof for classical number theory in which induction was applied only to quantifier free formulæ.

In the same year Ackermann improved his 1924 proof to give consistency for the same system as von Neumann. It seemed that, with a relatively straightforward extension of these arguments, it would soon be possible to prove number theory consistent. But these hopes were to be dashed on 7 September 1930 by Gödel's announcement in Königsberg of the first results of his current research, including his first incompleteness theorem. Although Gödel had only proved his first incompleteness theorem at this stage, his theorem already contained the seeds of the destruction of Hilbert's program in that any possible consistency proof for a system containing number theory would now have to be for an incomplete system.

This theorem was subsequently published in "Über formal unentscheidbare Sätze der Principia mathematica und verwandter Systeme, I" along with his second major incompleteness theorem, on the impossibility of a proof of consistency formalizable within the system itself, that he proved a little while later. An abstract containing the main results for this paper was presented on 23 October 1930 to the Vienna Academy of Sciences by Hans Hahn, the paper itself being received for publication on 17 November 1930.[15]

By 1929 Gödel had read Hilbert and Ackermann's book, *Grundzüge der theoretischen Logik*, and had taken from it the open problem on completeness of the restricted predicate calculus. He solved this problem as his doctoral dissertation, completing it in the autumn of 1929. Gödel was familiar with Hilbert's consistency program and in 1930 had begun to consider the consistency of analysis. At this point we will quote extensively from Hao Wang, who discussed these issues with Gödel before Gödel's death in 1978, and who reported on them in his article entitled, "Kurt Gödel's Intellectual Development."[16]

> In the summer of 1930, Gödel began to study the problem of proving the consistency of analysis. He found it mysterious

---

[15]See [**169**, p. 654 − 655] for details of the chronology.
[16]See [**168**, p. 183-184].

that Hilbert wanted to prove directly the consistency of analysis by finitist methods. He believed generally that one should divide the difficulties so that each part can be overcome more easily. In this particular case, his idea was to prove the consistency of number theory by finitist number theory, and prove the consistency of analysis by number theory, where one can assume the truth of number theory, not only the consistency. The problem he set for himself at that time was the relative consistency of analysis to number theory; this problem is independent of the somewhat indefinite concept of finitist number theory.

He represented real numbers by formulas, or rather propositional functions, of number theory and found he had to use the concept of truth for sentences in number theory in order to verify the comprehension axiom for analysis. He quickly ran into the paradoxes (in particular, the Liar and Richard's paradox) connected with truth and definability. He realized that truth in number theory cannot be defined in number theory and therefore his plan of proving the relative consistency of analysis did not work. He went on to draw the conclusion that in suitably strong systems such as certain small subsystems of type theory or set theory, there are undecidable propositions.

So Gödel's theorems, in part at least, can be seen as a direct outcome of Hilbert's program itself. Hilbert, himself, still retained some hope for his plan and in 1936 Gentzen found a consistency proof for classical number theory that used transfinite induction up to ϵ_0, the limit of the sequence ω, ω^ω, ω^{ω^ω}, ... This was a step that was never envisaged by Hilbert when he first instigated the search for consistency proofs.

3.6. Gödel's 1931 Paper

Gödel's 1931 paper is a landmark in the development of modern logic. It has been translated, studied and analyzed many times by many scholars.[17] The paper was of major importance to the Hilbert school and to logic in general but it is also of importance in its use of recursive functions, and the impetus it gave to the study of function definitions, mechanical procedures, formal systems and to decidability problems. Gödel's writings are always clear and precise and his work on recursive functions is no exception to that rule.

There are four sections to Gödel's paper, his two main theorems are contained in Sections 2 and 4. The first section includes an informal sketch of the main arguments and results which he derives later, along with a note on the relationship between his arguments and the Liar and Richard paradoxes. The whole of Gödel's argument centers around the Liar paradox, with truth replaced by provability in the formal system involved. The paradox itself is not operative in this situation but Gödel proves that there exist unprovable formulae in the formal system which are true under the standard interpretation.

The second section contains the main theorem on the incompleteness of any formal system rich enough to express arithmetic. Before proving this main theorem Gödel arithmetizes the metamathematics and then proves that the resulting recursive functions and relations can be expressed within the system itself (Theorem V). Since Gödel is intending to prove a result about the completeness of a formal system for arithmetic he starts by describing that formal system. This system is essentially the logical axioms of *Principia Mathematica* plus the arithmetic axioms of Peano with the natural numbers as variables of type one (individuals). This is his basic system, which be calls P, though later

---

[17]See for instance De Long [41], Hilbert and Bernays[81], Kleene [91], Kneebone [98], Nagel and Newman [109], and Rosser [140].

he fortifies it for various theorems and considers its general properties in order to extend his results.

In itself, of course, the formal system was not new, but what Gödel realized was that this structure could be coded into natural numbers and he specifically designed this coding so that the rules which govern the formation of formulae and the rules which govern the proof of formulae and so on could be expressed in terms of numerical relationships between the representative natural numbers. This arithmetization of the formal system is now called Gödel numbering.

With each primitive sign, with each sequence of primitive signs called formulæ and with each sequence of formulæ Gödel associates a natural number. He denotes by $\Phi(a)$ the number that is assigned to the primitive sign or sequence of primitive signs a.

For any relation between primitive signs or sequence of signs $R(a_1, \ldots, a_n)$, he associates the relation $R'(x_1, \ldots, x_n)$ between natural numbers $x_1, \ldots, x_n$ that holds if and only if $R(a_1, \ldots, a_n)$ holds and $x_i = \Phi(a_i)$ for $i = 1, \ldots, n$.

In this way he can translate metamathematical assertions about the theory into numerical relationships between natural numbers coding various parts of the theory. So that, for example, $y = \text{Neg}(x)$ is a relationship between two natural numbers that holds if and only if y is the Gödel number of a formula a_1 and x is the Gödel number of a formula a_2 and $a_1 = \neg(a_2)$.

The idea of Gödel numbering can be seen as a natural development from Hilbert's treatment of formulae as meaningless signs. As Gödel says:[18]

> Of course, for metamathematical considerations it does not matter what objects are chosen as primitive signs, and we shall assign natural numbers to this use.

_____

[18]See [**64**, p. 597].

The innovation of using numbers for primitive signs was taken by Gödel, but the real power behind the idea was Gödel's effective use of it in treating metamathematical arguments numerically. The concept of Gödel numbering represents a very significant development and its usefulness and power have led to it becoming a common tool in foundational research.[19]

In this second section of Gödel's paper he produces forty-six numerical relations and functions, most of which express metamathematical notions. For example:

    i) Relation number 45 is xBy which says "x is the PROOF of a FORMULA y" and is a numerical relationship which holds between two numbers x and y if and only if x is the Gödel number of a proof of the formula with Gödel number y, and

    ii) Relation number 46 is $Bew\ (x)$ which says that "x is a PROVABLE FORMULA" that is, x is the Gödel number of a provable formula.

This arithmetization of the metamathematics gains its full power from the fact that Gödel proves that it is possible to formalize the metamathematics of P within P itself. It is then possible to argue about the consistency of P or provability in P inside P itself via this formalized arithmetization of the metamathemtics. According to Nagel and Newman the importance of this technique was comparable in fertility to Descartes' introduction of the algebraic method into geometry, though this probably overstates the case a little.

The proof that these numerical relationships can be formalized in P is accomplished by Gödel in two stages. He first shows that all of the numerical

---

[19]The idea of arithmetizing the metamathematics was also independently conceived by Tarski in "The concept of truth in formalized languages" (See page 184 of this English translation contained in *Logic, Semantics, Metamathematics* [**159**]) though, by Tarski's own admission, Gödel's development was much more complete. Tarski's original paper was researched in the period 1929-31 but not published until 1933.

relationships except the last, number 46, are primitive recursive.[20] He then states his fundamental Theorem V which basically expresses the fact that for every recursive number-theoretic relation there is a formula in the system P such that this formula "numeralwise expresses" the given relation. This terminology is used by Kleene in his *Introduction to Metamathematics*[**91**]. This theorem is crucial to Gödel's argument and it says that if $R(x_1, \ldots, x_n)$ is a recursive relation then there exists an expression in P such that whenever the recursive relation R is true for particular numbers then its corresponding expression in P is provable and whenever the recursive relation is false the negation of its corresponding expression in P is provable. The intention of this correspondence theorem of Gödel's is that the recursive relations will express the expected interpretation of the symbolism of P. We therefore have that provability in P is the syntactical counterpart of the semantic notion of recursive truth.

This theorem is not proved in Gödel's paper, but it is argued that the necessary proof, while long winded, would not present any great difficulties. Besides the formal system P was set up in the first place with the express purpose of including formulae in it which are provable if and only if some recursive arithmetic relations are true.

Recursive relations and functions thus play a major role in Gödel's paper. We will therefore describe his definitions of these functions. He starts by giving the format for definition of a function by recursion, namely:[21]

A number-theoretic function $\phi(x_1, \ldots, x_n)$ is said to be re-cursively defined in terms of the number-theoretic functions

---

[20]The term "primitive recursive function" was not used until later. For the rest of this chapter we will use "recursive function" for "primitive recursive function" unless otherwise indicated.

[21]See [**64**, p. 602].

$\psi(x_1, \ldots, x_n)$ and $\mu(x_1, \ldots, x_n)$ if

$$\phi(0, x_2, \ldots, x_n) = \psi(x_2, \ldots, x_n)$$

$$\phi(k + 1, x_2, \ldots, x_n) = \mu(k, \phi(k, x_2, \ldots, x_n), x_2, \ldots, x_n)$$

hold for all $x_2, \ldots, x_n, k$.

This schema for definition by recursion is first given by Gödel in this paper and is the one that is generally adopted today. As has been noted previously in this chapter, it was proved equivalent to Hilbert's schema by Péter in 1934.

He then defines what is needed for a function to be recursive:[22]

> A number-theoretic function ϕ is said to be *recursive* if there is a finite sequence of number-theoretic functions $\phi_1, \ldots, \phi_n$ that ends with ϕ and has the property that every function ϕ_k of the sequence is recursively defined in terms of two of the preceding functions, or results from any of the preceding functions by substitution, or, finally, is a constant or the successor function $x + 1$. The length of the shortest sequence of ϕ_i corresponding to a recursive function ϕ is called its *degree*.

This is the first time that such an explicit statement is made about the class of recursive functions. It is a precise statement of something which had generally been practised before but without the limited number of initial functions necessary having been clearly expressed. Gödel only specifies two initial functions in this statement, the constant and successor functions, but recognizes later in 1934 that in fact a third initial function, the identity function, is advisable to give a complete, explicit definition.

---

[22]See [**64**, p. 602].

A relation $R(x_1, \ldots, x_n)$ between natural numbers is defined to be recursive if there is a recursive function $\phi(x_1, \ldots, x_n)$ such that, for all $x_1, \ldots, x_n$

$$R(x_1, \ldots, x_n) \text{ holds if and only if } \phi(x_1, \ldots, x_n) = 0.$$

This is the first definition of recursive relations apart from Skolem's awkward definition of 1923 and it has since become the standard definition. In particular it allows Gödel to define a recursive class as a one-place recursive relation.

Having thus defined recursive functions and relations, Gödel goes on to prove four theorems concerning recursive relations. In the last of these theorems he states that:

> If the function $\phi(x_1, \ldots, x_n)$ and the relation $R(x, y_1, \ldots, y_m)$ are recursive then so is the function $\psi(x_1, \ldots, x_n, y_1, \ldots, y_m)$ defined by
>
> $$\psi(x_1, \ldots, x_n, y_1, \ldots, y_m) =$$
> $$\epsilon_x[x \leq \phi(x_1, \ldots, x_n) \ \& \ R(x, y_1, \ldots, y_m)]$$
>
> where $\epsilon_x F(x)$ means the least x for which $F(x)$ holds, and zero in the case that there is no such number.

As we shall see in a later chapter, the bound on the value of x given in the above example makes ψ primitive recursive whereas without it ψ would have been general recursive.

Having used recursive functions and relations in the important task of representing the metamathematics of P in P itself, Gödel is ready to consider his main theorem, Theorem VI. Before proving this theorem Gödel fortifies his system P by adding to the axioms of P a set of formulæ with Gödel numbers in a class κ. He places two restrictions on the formulæ admissible in this class:

First, the set of consequences of the formulæ and the axioms
of P should be ω-consistent.

By this he means that the set of consequences should not contain some formulæ
$P(x)$ such that

$$P(0), P(1), P(2), \ldots \text{ and } \neg((\forall x)(P(x)))$$

are also contained in that set of consequences. For a formal system, if there
exists some formula $P(x)$ such that it is possible to prove

$$\neg((\forall x)(P(x))) \text{ and } P(0) \text{ and } P(1) \text{ and } P(2) \text{ and } \ldots$$

then that formal system is ω-inconsistent. The concept of ω-consistency was
first used by Tarski, though be did not name it that.[23] ω-consistency is stronger
than consistency since ω-consistency implies consistency, but it is possible for
a system to be consistent without being ω-consistent. The argument used in
Gödel's main theorem can be used to prove this result.

Second, the class κ should be recursive, that is, the set of Gödel
numbers of the extra formulæ should be recursively defined.

Gödel then states and proves his Theorem VI. In this theorem he constructs
a formula of the system P of the form $(\forall x)(A(x))$ whose intended interpretation
represents the metamathematical statement that the formula itself is unprovable
in the system involved. This formula corresponds to the Liar Paradox but with
truth replaced by provability. In the first part of the proof he shows that this
constructed formula is provable from the axioms and added formulae given by
κ only if the system $P + \kappa$ is inconsistent and therefore ω-inconsistent. He then

---

[23]Tarski used the concept of ω-consistency in 1927 at the second conference of the Polish
Philosophical Society in Warsaw. This led to the publication in 1933 of the article, "Einige
Betrachtungen über die Begriffe ω-Widerepruchsfreiheit und der ω-Vollstdhdigkeit" (Some
observations on the concepts of ω-consistency and ω-completeness),though by then he had
seen Gödel's paper and consequently used Gödel's name for the concept.

shows that the assumption that the negation of this formula can be proved from the axioms and formalæ given by κ will lead to the construction of an ω-inconsistency. Hence if the system $P + \kappa$ is to be ω-consistent then there must exist within the system a formally undecidable formula, that is, a formula such that neither it nor its negation is provable within the system.

But the formula that Gödel constructs has an intended interpretation which represents the metamathematical statement that the formula is unprovable, and as we have seen from the first part of the proof of Theorem VI this is, in fact, true. Hence the formula that is undecidable in the system $P + \kappa$ is decidable metamathematically and is in fact true. So Gödel has shown the system $P + \kappa$ to be incomplete in that it contains a formula such that both it and its negation are unprovable but which is supposed to represent a formula that can be shown to be true by other methods.

After his proof of this theorem Gödel makes four rather important comments.

First, he states that if $P + \kappa$ is assumed consistent, rather than ω-consistent, then the existence of an undecidable formula does not follow from the previous argument.

Rosser did, however, succeed in extending Gödel's result for $P + \kappa$ assumed only consistent by constructing a more complicated formula.[24]

Second he comments that if the negation of his constructed formula $\neg(\forall x)(A(x))$ is added to the axioms to form a system $P + \kappa'$ then that system must be consistent or else the original formula $(\forall x)(A(x))$ would have been actually provable in $P + \kappa$ itself. But, by the second part of his theorem, it is also ω-inconsistent.

---

[24]See [**138**].

That is, Gödel shows $A(0)$, $A(1)$, $A(2)$,... are all provable while $\neg(\forall x)(A(x))$ is also provable by assumption. This shows that a system can be consistent and yet ω-inconsistent at the same time on the assumption that some $P + \kappa$ is consistent in the first place.

A system in which for some $A(x)$ it is possible to prove $A(0)$, $A(1)$, $A(2)$,... for all x and not prove $(\forall x)(A(x))$ is now called ω-incomplete.[25]

Third, he defines a decidable relation. The number-theoretic relation $R(x_1, \ldots, x_n)$ is decidable if there exists a relation r in P such that whenever $R(x_1, \ldots, x_n)$ is true r is provable in P and whenever $R(x_1, \ldots, x_n)$ is false $\neg r$ is provable in P. Here the numerals that represent the natural numbers x_i, are substituted for the variables in r. That is, a number-theoretic relation is decidable if it is the intended interpretation of some provable formula in P. Since this is true for all recursive relations then all recursive relations are decidable relative to P.

Fourth and finally, he observes that in the whole of his proof of Theorem VI he only uses two properties of P: (1) that the class of axioms and rules of inference are recursively defined and (2) that every recursive relation is definable in the system P in the sense of Theorem V.

From this he concludes that any other ω-consistent formal system that satisfies (1) and (2) will have in it undecidable propositions of the form $(\forall x)(F(x))$, where F is a recursively defined property of natural numbers. In particular this has the very important consequence that the Zermelo-Fraenkel set theory and von Neumann's set theory are formal systems to which Gödel's Theorem VI apply, that is, they are both incomplete.

In Section 3 of his paper Gödel uses the results of Theorem VI to produce two further undecidability theorems. Here we will just state the two results and

---

[25][**157**] and see footnote 23.

also the definition of an arithmetic relation, which is necessary for the first of these two theorems.

> A relation is said to be *arithmetic* if it can be defined in terms of the notions of addition and multiplication for natural numbers and the logical constants $\vee$, $^-$, (x), and $=$, where (x) and $=$ apply to natural numbers only.[26]

Gödel then states Theorem VII:

> Every recursive relation is arithmetic.

His proof makes use of the way recursive functions are derived in order from the initial functions, that is, he uses induction on the degree of a recursive function.

From this he deduces the first undecidability results of this section, Theorem VIII, which states that:

> In any formal system mentioned in Theorem VI there are undecidable arithmetic propositions.

Finally he gives the second undecidability result which shows that in any formal system formed by fortifying P as in Theorem VI, there are undecidable satisfiability problems of the restricted functional calculus of Hilbert and Ackermann [79]. He does this by producing a formula of the restricted functional calculus whose satisfiability is dependent on whether or not $(\forall x)(F(x))$ holds for some recursive F, and then utilizes the fact that there are undecidable propositions of the form $(\forall x)(F(x))$ in any of these formal systems.[27]

---

[26]The notation $^-$ is the symbol for "not" and (x) is read "for all x". See [**64**, p. 610].

[27]See the fourth comment after Theorem VI.

Section 4 consists of only one theorem, Theorem XI, but this theorem is an extremely important consequence of Gödel's main theorem and is the second major result in his paper.

Theorem XI concerns consistency. In the first part of the proof of Theorem VI Gödel shows that:

> If the system $P+\kappa$ is consistent then some constructed formula, call it G, is unprovable.

In this new theorem, Gödel first uses the example "there exists an unprovable formula in the system" as a representative of a metamathematical statement that expresses the consistency of $P + \kappa$. He then suggests how this statement can be coded via Gödel numbers into a number-theoretic relation, $Wid(\kappa)$. This arithmetic relation is the intended interpretation of some formula W in $P + \kappa$. Gödel shows informally that the formula $W \to G$ should be provable in $P + \kappa$ so that, if W is provable in $P + \kappa$ then, by *modus ponens*, so would G be provable. But by Theorem VI the constructed formula G is not provable in $P + \kappa$, providing $P + \kappa$ is consistent, so W must not be provable in $P + \kappa$.

Hence Gödel has shown that the formula W in $P + \kappa$, which has as its intended interpretation a numerical relation expressing the consistency of that system, is not provable in that system itself, providing the system is in fact consistent. That is, there can be no consistency proof of the system which is formalizable within that system.

Gödel intended to give a detailed formal proof of this result in a further paper but found it unnecessary to do so since his results were rapidly accepted.[28]

---

[28]The detailed proofs can be found in Hilbert and Bernay's *Grundlagen der Mathematik* Volume 2.

Gödel completes his paper with two comments on this final theorem. First, he states that his final theorem also applies to set theory (M) and to classical mathematics (A): [29]

> There is no consistency proof for M, or for A, that, could be formalized in M, or A, respectively, provided M, or A, is consistent.

Second, he makes the following observation on how his results affect Hilbert's plan.[30]

> I wish to note expressly, that Theorem XI (and the corresponding results for M and A) do not contradict Hilbert's formalistic viewpoint. For this viewpoint presupposes only the existence of a consistency proof in which nothing but finitary means of proof is used, and it is conceivable that there exist finitary proofs that *cannot* be expressed in the formalism of P (or of M or A).

3.7. Comments and Conclusion

Gödel's 1931 paper, in common with all of his papers, is characterized by precision and thoroughness of detail. This paper is important not only for the results it presents but also for the rich fund of new ideas and methods that it introduces. One particular method, that of arithmetization of the metamathematics, has become an everyday tool in foundational research. The results and, methods contained in this paper provided a stimulus for the study of many foundational questions, including questions of decidability and of effective methods of procedure.

---

[29]See [**64**, p. 615].
[30]Ibid.

The two fundamental results pertaining to foundational research contained in this paper are:

> For certain formal systems intended to contain arithmetic and obeying certain restrictions, 1) these systems, if ω-consistent, are not complete in that there exist formally undecidable formulae in them and 2) these systems, if consistent, are such that the proof of their consistency can not be formalized within them.

Despite Gödel's final comments, this paper caused Hilbert's plan to be abandoned in the form that Hilbert had previously envisaged it. Gödel was a very cautious person and in this paper he is concerned about the generality of his arguments and whether they applied to all formal systems so he is more cautious in his comments about Hilbert's plan than he needs to be. He was certainly quite convinced that Hilbert's program was invalidated after Turing's work on computable functions in 1936. As has already been said, many research workers gave up the attempt to prove arithmetic consistent. Those who did continue knew that they would have to use something more powerful than previously accepted finitary concepts, their task, however, being to minimize the use of these non-finitary concepts yet still prove consistency. Gentzen, in 1936, was the first to succeed in this task.

Following Gödel's paper, the concept of a formal system has been sharpened and consequently Gödel's theorems have been generalized and applied to any system containing at least primitive recursive arithmetic. In a paper [52] written by Gödel fairly soon after this main paper there is a result showing that, if successive enlargements are made to the system Z by adding higher type variables, then a sequence of formal systems can be obtained in which undecidable propositions of earlier systems become decidable, while new undecidable propositions are constructed. The system Z is the formal system that contains the Peano axioms, the schema for definition by recursion on one variable, the logical rules of the restricted functional calculus and contains no variables other than variables for individuals.[31].

The proofs of Gödel's theorems have subsequently been shortened and improved and, as mentioned earlier, result (1) no longer requires the assumption of ω-consistency but only consistency.

Probably the first published research in this field following Gödel's paper was Herbrand's last paper [**67**]. Although most of this paper was written before Herbrand knew of Gödel's theorems, and although it contains a proof of the consistency of a part of arithmetic, Herbrand did have time to read Gödel's results and comment on them before he sent his paper for publication. He gives a brief proof of Gödel's Theorems VI and XI and then discusses the reasons why they do not apply to the part of arithmetic on which be is working.

His main argument is that his own system contains a set of function-defining conditions that preclude the possibility of the enumeration of all the functions and so it is impossible to express the metamathematics of the system within the system as required by Gödel. As we shall see later, this set of conditions is rather important and represents a forerunner of the definition of general recursive functions.

Herbrand completes his paper by commenting on how Gödel's results apply to Hilbert's plan:[32]

> It seems to us impossible, contrary to Gödel's opinion, that there could be intuitionistic [meaning finitary in this case] arguments not formalizable in ordinary analysis.

---

[31]See [**81**] for results on other formal systems
[32]See [**64**, p. 628].

Gödel used primitive recursive functions and relations as the cornerstone for his rigorous proof of his incompleteness theorem and thus needed to considerably improve the theory of recursive functions. In so doing he produced many important new results. He presented a general schema for the definition of recursive functions that was more general than any given before and made more precise the class of functions involved. He was specific about his use of initial functions and he gave the definition of recursive relations that has been used ever since.

In his definition of recursive functions he remained very much with the spirit of Skolem's work, that is, he used:

  i) the constant function which is Skolem's natural numbers
  ii) the successor function exactly the same Skolem's
  iii) a definition by substitution which is Skolem's substitution of equals for equals
  iv) a definition by recursion directly analogous to Skolem's recursive mode of thought.

Later, in 1914, Gödel added the identity function but even this was still consistent with Skolem's early work.

Finally amongst these numerous results on recursive functions Gödel solved the problem of eliminating bounded quantifiers from recursive definitions. In Theorem IV he shows how the two recursive relations, S and T, and a recursive function ψ

$$S(\mathbf{x}, \mathbf{y}) = (\exists x)[x \leq \phi(\mathbf{x}) \;\&\; R(x, \mathbf{y})]$$

$$T(\mathbf{x}, \mathbf{y}) = (\forall x)[x \leq \phi(\mathbf{x}) \;\&\; R(x, \mathbf{y})]$$

$$\psi(\mathbf{x}, \mathbf{y}) = \epsilon_x[x \leq \phi(\mathbf{x}) \;\&\; R(x, \mathbf{y})]$$

defined originally in all generality in terms of bounded quantifiers the recursive function $\phi(\mathbf{x})$, and the recursive relation $R(x, \mathbf{y})$, can also be defined using only his standard schema for recursive definition and previously defined recursive functions.[33]

---

[33]$\mathbf{x}$ and $\mathbf{y}$ are abbreviations for arbitrary n-tuples of variables.

CHAPTER 4

Early Work Leading to λ-Definable Functions

4.1. Introduction

Princeton, in the early 1930's, was one of the main centers of research from which today's theory of recursive functions was to spring. The initial inspiration and guidance was given by Alonzo Church, who was Assistant Professor in Princeton from 1929 until 1939 when he was made an Associate Professor. He initiated the early work by lecturing on his current research topic which he had commenced in 1928 while he was a National Research Fellow. Church was attempting to produce a set of postulates for the foundation of logic in which he hoped to develop the theory of positive integers. In this task he was aided by two of his research students, Stephen C. Kleene and J. Barkley Rosser. Kleene worked directly on the theory of positive integers within Church's system of logic while Rosser worked on similar topics in an associated system of logic developed by Curry. Their work was closely interwoven and they made extensive use of each other's results.

All of their papers belong to the period after Gödel's 1931 incompleteness theorem, but Church's first, paper was sent for publication before he knew of Gödel's result. In fact Kleene recalls that Church had already made available a draft of his second paper, which was being discussed in lectures, before they were acquainted with Gödel's theorems. Church was interested in proving the consistency of his set of postulates, this task becoming even more important once Gödel's results were known. He came tantalisingly close to success. Just one case of induction failed. But in the autumn of 1933 Rosser came to suspect that an inconsistency existed in Church's system and he proposed a plan for the proof of such an inconsistency. He informed Church of his suspicions and, in December 1933 or January 1934, Kleene was recruited to help because of his expertise in dealing with Church's system of logic.

By about March or April of that year they were fairly sure that the proposed proof would be successful and the final paper containing this proof of the inconsistency was received for publication on 13 November 1935. After Rosser had informed Church of his plan for proving the inconsistency of the full system of logic Church began to stress the significance of the function defining structure that could be abstracted from this system. This function defining structure was originally known as formal definability, but due to Church's use of λ to represent the function it later became known as λ-definability. When Church first started working with his system he had no idea how many functions would prove to be λ-definable but as each new function that was tried proved to be λ-definable Church began to speculate that perhaps all effectively calculable functions were λ-definable. Church finally proposed this thesis equating the classes of λ-definable functions and effectively calculable functions early in 1934. Subsequently Gödel introduced him to another concept, that of general recursive functions, so that when Church published his thesis in 1936 he included both the classes of λ-definable functions and general recursive functions as being equivalent to the class of effectively calculable functions.

In this chapter we shall see the development of the class of λ-definable functions and see the introduction of some of the essential theory needed to prove the equivalence between the aforementioned classes of functions.

4.2. Church's System of Logic

Church's first paper was intended as the first stage in founding a new system of formal logic. His approach shows some influence of Hilbert's style due in part, no doubt, to Church having been in Göttingen from 1928 to 1929 while the first draft of his work was being prepared. Church hoped that his system of logic would be adequate for the development of mathematics, but he was rather hesitant about claiming this and also about claiming that his system was free from contradiction. He admitted that these questions were ones that he had not yet had time to investigate but hoped that further work would yield answers to these questions and perhaps indicate any modifications that would become necessary.

His treatment is formal but he had in mind an intended application for his logic which was, in fact, his motive for constructing the logic, so that while he defines his terms formally he also gives their intended meanings.

For instance, he specifies certain undefined symbols:[1]

$$\{\,\}, (\,), \lambda[\,], \Pi, \Sigma, \&, \sim, \iota, A$$

and symbols for variables. He then defines inductively the phrase well-formed formulæ including

> If $\mathbf{x}$ is a variable and $\mathbf{M}$ is well-formed then $\lambda\mathbf{x}[\mathbf{M}]$ is well-formed.

He then explains what he means by his λ symbol:[2]

---

[1]See [**17**, p. 351–352].

[2]*Ibid.* p. 352

> If **M** is any formula containing the variable **x**, then $\lambda\mathbf{x}[\mathbf{M}]$ is
> a symbol for the function whose values are those given by the
> formula.

In other words $\lambda\mathbf{x}[\mathbf{M}]$ represents the function **M** of one variable **x** rather than the ambiguous value of **M** evaluated at **x**. The variable **x** in the formula $\lambda\mathbf{x}[\mathbf{M}]$ is called a bound variable; any other variables that might occur in **M** are called free variables. Church's λ-notation is new to this paper and represents a useful way of distinguishing between a function value, a function of one variable or a function of two variables and so on. This λ-notation becomes, in later papers, an essential part of the machinery for defining functions which are then called λ-definable.

All the other undefined symbols are also given intended meanings.

Church then gives his rather extensive list of postulates, which includes thirty-seven axioms and five rules of procedure. Church uses the terminology "is true" to mean that a certain formula belongs to his abstract system - rather than the, by then, more usual and consistent term "is provable", which leaves the words "is true" to refer to the intended model.

From a modern perspective the first three rules of procedure are really the most important part of this paper and they are basically all that have survived. If **M** and **N** are well formed formulae and **N** can be derived from **M** by successive applications of only these three rules then **M** is said to be convertible into **N** and the process is spoken of as a conversion of **M** into **N** and written **M** conv **N**. As it is these three rules that eventually become the basis for λ-definable functions, we will quote each of them in full before discussing them. The symbol $S_{\mathbf{Y}}^{\mathbf{X}}\mathbf{U}|$ that appears in them stands for the result of replacing **X** by **Y** throughout the formula **U**, where **Y** may be any symbol or formula and **X** must be a single symbol.[3]

---

[3]Kleene later extends this so that **X** can also be a formula.

Rule I: If **J** is true, if **L** is well-formed, if all the occurrences of the variable **x** in **L** are occurrences as a bound variable, and if the variable **y** does not occur in **L**, then **K**, the result of substituting $S_{\mathbf{y}}^{\mathbf{x}}\mathbf{L}|$ for a particular occurrence of **L** in **J**, is also true.[4]

This rule essentially allows a change of variable in part of a formula. There are various restrictions on its use, in particular the original variable **x** must be bound in the part of the formula under consideration and the introduced variable **y** must not be already in the formula. These restrictions are designed to ensure that the resulting formula is still well-formed. Thus Rule I changes variables.

Rule II: If **J** is true, if **M** and **N** are well-formed, if the variable **x** occurs in **M**, and if the bound variables in **M** are distinct both from the variable **x** and from the free variables in **N**, then **K**, the result of substituting $S_{\mathbf{N}}^{\mathbf{x}}\mathbf{M}|$ for a particular occurrence of $\{\lambda\mathbf{x}\cdot\mathbf{M}\}(\mathbf{N})$ in **J**, is also true.[5]

This rule partly evaluates a function, that is, if **N** is the formula for a number and **x** is a variable then the function **M** of **x**, **M**(...**x**...), becomes **M**(...**N**...). Again restrictions are necessary to maintain well-formed formulæ.

Thus Rule II substitutes the argument of a function into the function.

Rule III: If **J** is true, if **M** and **N** are well-formed, if the variable **x** occurs in **M**, and if the bound variables in **M** are distinct both from the variable **x** and from the free variables in **N**,

---

[4]See [**17**, p. 355].
[5]See [**17**, p. 355].

then **K**, the result of substituting $\{\lambda\mathbf{x}\cdot\mathbf{M}\}(\mathbf{N})$ for a particular occurrence of $S\,_{\mathbf{N}}^{\mathbf{x}}\mathbf{M}|$ in **J**, is also true.[6]

Rule III does the opposite to Rule II. That is, Rule III reverses the substitution of the argument of the function.

The three rules together merely rewrite well-formed formulæ, as the name "conversion" indicates. They represent an abstract treatment of the notion function of an argument and, as will be seen later, these three rules, together with a few other definitions, were to lead to a very powerful function defining apparatus.

Having set up his formal system the rest of Church's paper is devoted to proving three metamathematical theorems. One of the innovations introduced in this paper is Church's restriction on the law of the excluded middle. He recognizes that any system of logic must be capable of avoiding the paradoxes of mathematics associated with the transfinite, but does not want to use the methods of either Whitehead and Russell or Zermelo which he considers rather artificial.

He proposes, therefore, to restrict the law of the excluded middle so that a propositional function **F** may, for some value **X** of the independent variable, represent neither a true proposition nor a false proposition. Thus, **F(X)** is undefined and represents nothing. His system of logic is therefore designed to deal with functions with limited ranges of definition, For example $\Pi(\mathbf{F},\mathbf{G})$ is intended to mean "$\mathbf{G}(x)$ is a true proposition for all values of x for which $\mathbf{F}(x)$ is a true proposition".[7]

Because he is replacing the law of the excluded middle by a weaker assumption, Church briefly considers the relationship between his system and

---

[6]*Ibid.* p. 356.
[7]See [**17**, p. 353].

Intuitionism. Both systems reject part of the principle of the excluded middle but the parts that are rejected are different for each system. For instance Brouwer rejects the law of double negation whereas Church accepts it in Postulates 26 and 27. Brouwer accepts that a statement from which a contradiction can be inferred is false while Church allows that statement to be undefined in some cases.

In particular Church's restriction allows him to avoid the Russell paradox since, although the formula, $\mathbf{P}(= \{\mathbf{M}\}(\mathbf{M}))$ that leads to the paradox can be written in Church's system and although assuming $\sim \mathbf{P}$ allows $\mathbf{P}$ to be inferred and vice versa, this does not lead to a contradiction since $\mathbf{M}$ need be neither true nor false for the value $\mathbf{M}$ of its independent variable.[8]

As already indicated, this first paper of Church, although it appears in the 1932 volume of *Annals of Mathematics* belongs to the period prior to Gödel's incompleteness theorem. The paper containing Gödel's famous theorem was known in Europe in 1930, since part of the results to be contained in that paper were announced at a conference in Königsberg on 7 September 1930 and an abstract of the paper was presented to the Vienna Academy of Sciences by Hans Hahn on 23 October 1930. But Church, working in Princeton, apparently did not know of Gödel at the time of presenting his paper for publication on 5 October 1931.

The reasons for believing this are threefold. First, Church claims, in a footnote on the first page, that the paper contains work, in a revised form, of the author while a National Research Fellow in 1928-29 [in Göttingen].

Secondly, Church discusses the consistency of his system without any reference to Gödel, whereas a reference to Gödel would seem imperative at this point.

---

[8]The symbol $\sim$ meant "not" in Church's intended application.

Finally Kleene states early in *Reminiscences of Logicians* [9] that be believes that Church had not even heard of Gödel, since he did not teach Gödel's 1930 result[10] in his lecture course at Princeton in the autumn of 1931. Furthermore, Kleene states, in a letter to the author of 28 July 1977, that he does not believe the second of Church's papers [18] could have been significantly influenced by Gödel's paper of 1931 either since he remembers a manuscript version of at least part of Church's second paper being available in class before Church heard of Gödel's 1931 result that autumn.

One of those present at the conference at Königsberg, where Gödel first announced some of his 1931 results, was von Neumann. Von Neumann was Visiting Professor of Mathematical Physics at Princeton in 1930, and Professor of Mathematical Physics from 1931 to 1933. In the autumn of 1931 be spoke at a mathematics colloquium. Instead of talking of his own, not inconsiderable, work he surprised those present by giving a lecture on Gödel's work. It was at this lecture that Church and Kleene first heard of Gödel's incompleteness result and Church realized at once that it posed a threat to his system of logic. But Church subsequently convinced himself that his system was sufficiently different in its formulation to make the theorem that the consistency of a system could not be proved within the system itself inapplicable. He makes a comment to this effect in his second paper.[11]

> The impossibility of such a proof of freedom from contradic-
> tion for the system of *Principia Mathematica* has recently been
> established by Kurt Gödel. His argument, however, makes use
> of the relation of implication U between propositions in a way
> which would not be permissible under the system of this pa-
> per, and there is no obvious way of modifying the argument so

---

[9]See [29, p. 2].

[10]"The Completeness of the Axioms of the Functional Calculus of Logic" [49] as trans-lated in [64, p. 582–591].

[11]See [18, p. 842–843].

as to make it apply to the system of this paper. It therefore remains, at least for the present, conceivable that there should be found a proof of freedom from contradiction for our system.

In Church's second paper, where he continues the process that he started in his first paper, he finds that he needs to modify one or two of his postulates to guarantee consistency. This paper does not contain very much of lasting importance, the most significant part in retrospect being his λ-definition of the positive integers. This definition of the positive integers is achieved by defining

$$1 \text{ as } \lambda f \lambda x . f(x)$$
$$S \text{ as } \lambda \rho \lambda f \lambda x . f(\rho(f, x))$$

and then denoting

$$S(1) \text{ by } 2$$
$$S(2) \text{ by } 3$$
.

so that S is the successor function. With these definitions it is found that

$$2 \text{ is convertible into } \lambda f \lambda x . f(f(x)),$$
$$3 \text{ is convertible into } \lambda f \lambda x . f(f(f(x)))$$

and so on. He gave these definitions because he still hoped to find that his system of logic was sufficient to develop the whole of mathematics. He therefore also indicated how to define some arithmetic operations and how Peano's axioms could be expressed in his system. Unfortunately Church found that the process of proving all Peano's axioms in his system and the task of extending this to rational and real numbers was far from easy. He therefore left this task as an open question at the end of his paper. As we see in the next papers, it was Kleene, his research student at Princeton, who took up this challenge when he was given, in February 1932, the task of developing the theory of positive integers in Church's system as a research problem for his doctorate.

4.3. The Development of Church's System

Kleene's first paper, "Proof by Cases in Formal Logic" [**84**], was submitted for publication in May 1933, five months after Church sent his own second paper for publication. Kleene's paper needless to say continues the work of Church.

Both Kleene's first paper and his subsequent PhD thesis paper, "A Theory of Positive Integers in Formal Logic" [**85**] were researched and written in the period January 1932 to June 1933. About October 1933 Rosser, who was also a research student at Princeton at the time, came to suspect that Church's system was inconsistent and by March or April 1934 Kleene and Rosser were just about convinced that their proof of the inconsistency would succeed. Consequently Kleene recalled both of his papers and revised them to contain, as far as possible, just that part of the theory of positive integers that, first, would be necessary to prove this inconsistency and, second, would include Church's λ-function defining apparatus that, by then, was seen to be of the most lasting interest.

In his 1934 paper Kleene starts by making some minor adjustments and improvements to Church's system of logic and proceeds to prove an important metamathematical theorem, the "Proof by Cases." In this paper Kleene intends to generalize on Church's work and wants to use as small a subset of Church's axioms as he can. He accomplishes this by not specifying as many undefined terms. Instead he defines a proper symbol as any symbol other than the following:

$$\{ \, , \ \}, \ (, \ ), \ \lambda, \ [\, , \ ]$$

so that he can introduce any other symbol he wants without remaining tied to Church's particular system of logic. He then improves Church's definition of well-formed formulæ calling them well-formed expressions and the laws relating to them, including the extension of $S_{\mathbf{Y}}^{\mathbf{X}} \mathbf{U}|$ to mean that $\mathbf{X}$ need not just be a variable but can be any well-formed expression. He accredits some of these theorem to Church's lectures at Princeton in November 1931.

In this paper Kleene uses the notion of combinations which he borrowed from the work of Rosser. Combinations were originally researched by Schönfinkel and Curry in the 1920's. In fact, Curry's theory of combinatorial logic was developed around 1930 as another independent system of logic much the same as Church's.[12] Curry's system was found to be inconsistent slightly later than was Church's system although both proofs were published at the same time. Rosser was working on bridging material between these two systems. Kleene made use of these combinations partly to provide convenient intermediate steps in proofs and partly to facilitate the borrowing of some theorems from Rosser. He also found them convenient to use in later, more important papers such as "λ-Definability and Recursiveness" [**87**].

By far the most important definition in Kleene's 1934 paper, though, is the definition of normal form which Kleene adopted from Church's lectures of autumn 1931.[13] This definition of normal form is:[14]

> $\mathbf{M}'$ shall be a normal form of $\mathbf{M}$, if $\mathbf{M}$ conv $\mathbf{M}'$ and $\mathbf{M}'$ contains no part of the form $\{\lambda\mathbf{x}.\mathbf{R}\}(\mathbf{S})$.

Hence "normal form" means that all uses of Rule II have been completed, that is, all substitutions for bound variables have been carried out. The concept of normal form furnishes a possible method of testing whether or not the formula $\mathbf{M}$ is convertible into the formula $\mathbf{N}$. First $\mathbf{M}$ and $\mathbf{N}$ are both converted into their normal forms $\mathbf{M}'$ and $\mathbf{N}'$, if such normal forms exist, and then $\mathbf{M}'$ and $\mathbf{N}'$ can be compared easily since only applications of Rule I are necessary to convert $\mathbf{M}'$ into $\mathbf{N}'$ if this is, in fact, possible. As we shall see later, in Church's paper,

---

[12]Both Curry and Church had spent time at Göttingen in 1928-1929.

[13]See [**87**, p. 341, footnote 8].

[14]See [**84**, p. 535].

"An Unsolvable Problem of Elementary Number Theory," [23] the process of being able to convert a formula into its normal form provides an algorithm for the effective calculation of a function of positive integers. Accordingly it becomes important to know whether or not such a function can be converted to its normal form or if such a function even has a normal form.

Kleene, in his paper, "λ-Definability and Recursiveness" recalls:[15]

> The notion of the *normal form* of a formula under conversion was originally introduced by Church in lectures at Princeton in the fall of 1931.

But Church states in the paper, "An Unsolvable Problem of Elementary Number Theory" cited above, that it was Kleene who thought of the problems of finding (1) an effective method of determining whether two formula **A** and **B** are such that **A** conv **B** and (2) an effective method of determining whether a formula **C** has a normal form, and that he proposed these problems to Church in conversation in about 1932.[16]

4.4. A Theory of Positive Integers in Formal Logic

The most explicit changes made to Church's system came in Kleene's two-part paper, "A Theory of Positive Integers in Formal Logic." This paper was originally submitted for publication in October 1933, but, as we have already noted, it was revised in April 1934 and resubmitted in June 1934 prior to publication in 1935. This paper is a detailed attempt to construct a significant portion of the theory of positive integers within a subset of Church's formal

---

[15]See [**87**, p. 341, footnote 8].

[16]See [**23**, p. 359, footnote 23]. Kleene was actually making some progress towards a solution of (1) about that time.

axioms. Kleene explicitly avoids the symbols for negation $\sim$, class A, and the description operator ι used by Church. Kleene observes:[17]

> ...significances can be assigned to the symbols of the logic in such a way that the formulas become assertions of logical truths. The mathematical interest arises from the fact that among the logical entities of the system it is possible to select certain ones which occur in relations of the same form as the relations between certain entities in mathematical theories. Hence if the mathematical entities are identified with, or defined to be, the logical entities, the propositions in which they occur will read as theorems of mathematics. It is in this sense that we are to develop or deduce the theory of positive integers within the logic.

The paper contains masses of detailed proofs on all manner of defined functions and, as such, is an extremely difficult paper to read. All the results are proved explicitly and after each definition all the properties that this abstract formula should possess in order to be identified with some mathematical function or relation are shown to follow from previous results. In subsequent papers much use is made of the results derived in this paper or results similar to them, so that, in these later papers, it is only necessary to state that the theorem is provable and refer the reader to Kleene's 1935 paper. Kleene's paper, therefore, contains all the details and ground-work for several later papers both from himself and from Church.

As noted, Church was originally somewhat dubious about the possibility of success for his plan to develop the theory of positive integers in his system. In particular, he had found no way to define the predecessor function needed for

---

[17]See [**85**, p. 153].

the third Peano axiom. Kleene reports in *Reminiscences of Logicians*[18] that he initially produced the predecessor function by changing Church's definition of formulae for integers. This meant that it was no longer possible to produce the recursive definitions that Church had carefully designed his system of integers to facilitate.

On being informed of this by Church, Kleene decided to abandon his new λ-definitions of the integers and continue with Church's formulation. As he says in a letter to the author of 3 November 1981, he had no proof at the time that it was not possible to produce alternate recursive definitions with these new λ-definitions, but that, since no way of doing this came immediately to mind, he went back to Church's identification of the integers and set to work to define the predecessor function.

He reports in *Reminiscences of Logicians* that the inspiration for the definition of the predecessor function came to him while he was at the dentist having two wisdom teeth out.[19] A little more work proved that this definition had the required properties. In his paper, "Origins of Recursive Function Theory," Kleene continues:[20]

> Thirty-one years later, in a letter to the author dated January 20, 1963, Dana Scott communicated to me an alternative identification of the positive integers with λ-formulas making the predecessor function immediate (the same as, or similar to, mine of 1932, of which I preserved no record), and indicated success in the further development of the theory using that alternative.

---

[18]See [**29**, p. 4].

[19]See [**29**, p. 5].

[20]See [**94**, p. 56].

In fact Kleene found that he was able to define in the system any function be wanted to define and Church began to speculate that perhaps all effectively calculable functions could be defined in his system. In his paper of 1935 Kleene succeeds in stating and proving the first, second, third and fifth Peano axioms; the fourth can no longer be stated since it involves negation. He then continues to show that proofs by induction, and some more generalized versions of induction, can be carried out in the system.

Kleene carefully redefines some of Church's definitions, such as + and ×, using some of Rosser's ideas so that the new definitions do not depend on Church's particular system of logic, and the more functions and relations he defines the more it became obvious that the functions themselves might be worthy of study in their own right.

As Kleene remarks on the first page of part two of this paper:[21]

> Closely connected with the formal theory of this paper, there
> is an intuitive theory concerning the formal definition of the
> functions involved.

Kleene had already noted the significance of the notion of formally definable functions, as abstracted from Church's system of logic, before his paper was revised in the light of the proof of an inconsistency in Church's system. It was, however, Church who first spoke out about the proposal to consider the function defining structure in its own right in response to Rosser's discovery of a probable contradiction in Church's system.

In his paper Kleene specifies what it means to formally define a function:[22]

---

[21]See [**85**, p. 219].
[22]See [**85**, p. 219].

> If L is an intuitive function which associates well-formed
> expressions $L(x_1, \ldots, x_n)$ with n-tuples $(x_1, \ldots, x_n)$ of well-
> formed expressions, then L shall be said to be defined for-
> mally by $\mathbf{L}$ if $\mathbf{L}(x_1, \ldots, x_n)$ conv $L(x_1, \ldots, x_n)$ for each set
> $(x_1, \ldots, x_n)$ for which L is defined.

In later papers functions defined this way are called λ-definable functions.

This definition can be illustrated by a simple example:

> If $\mathbf{x}$ and $\mathbf{y}$ are the formal equivalents in Church's notation of
> the arbitrary integers x and y, and if $+$ is a function defined in
> Church's system to be the addition function, then this function
> $+$ is formally defined if $\mathbf{x} + \mathbf{y}$ is convertible into $\mathbf{z}$ where $\mathbf{z}$ is the
> formal equivalent to z the intuitive sum of integers x and y.

This, then, is the first published recognition of Church's function defining apparatus abstracted from the rest of the logical structure. The remainder of Kleene's important paper is devoted to the further developments of formal definitions of functions and formal proofs in conjunction with each other. These further developments can be divided into three main sections which we shall discuss in the next three sections of this chapter.

4.5. λ-Definable Functions

While attempting to develop the theory of positive integers within Church's system of logic, Kleene became an expert in proving functions and operations to be formally definable. In Church's [23] and Kleene's [87] this is referred to as being λ-definable. In fact, once the predecessor function had been proved formally definable, in February 1932, it became a major subproject of Kleene's work to investigate other functions and operations and determine whether they too could be proved formally definable. It was recognized at the outset that only effectively calculable functions and operations that preserve effective calculability could be formally definable but just how many of them there were remained to be seen. As Kleene and Church thought up new examples of functions and operations so Kleene established the functions to be formally definable and the operations to preserve formal definability. It did not take long before the idea materialized that perhaps all effectively calculable functions were, in fact, formally definable.

The results that Kleene proved in his 1935 paper also eventually led to the proof of the equivalence between λ-definable functions, and general recursive functions, although at the time that this paper was first sent for publication Kleene was not aware of the general recursive functions of Herbrand and Gödel. He was, however, aware of Gödel's 1931 paper and so knew of the recursive functions, later called primitive recursive functions, contained in that paper.

In his paper Kleene proves that many primitive recursive functions are formally definable and considers the general schema for definition by primitive recursion. Thus, after a series of theorems concerned with formal definitions of sequences, he arrives at the following theorem:[23]

> If the free symbols of $\mathbf{F}$ are included in those of $\mathbf{A}$, then a formula $\mathbf{L}$ can be found such that
>
> $$\mathbf{L}(\mathbf{x}_1, \ldots, \mathbf{x}_n, \mathbf{y}_1, \ldots \mathbf{y}_m, \mathbf{1}) \text{ conv } \mathbf{A}(\mathbf{x}_1, \ldots, \mathbf{x}_n)$$
>
> and
>
> $$\mathbf{L}(\mathbf{x}_1, \ldots, \mathbf{x}_n, \mathbf{y}_1, \ldots \mathbf{y}_m, \mathbf{i} + \mathbf{1}) \text{ conv}$$
> $$\mathbf{F}(\mathbf{y}_1, \ldots, \mathbf{y}_m, \mathbf{i}, \mathbf{L}(\mathbf{x}_1, \ldots, \mathbf{x}_n, \mathbf{y}_1, \ldots \mathbf{y}_m, \mathbf{i}))$$
>
> for $x_1, \ldots, y_m = 1, 2, \ldots$.

---

[23]See [**85**, p. 223].

This result shows, with a slight adjustment concerning natural numbers and positive integers dealt with below, that the function L, defined by recursion from the functions A and F as given by Gödel in 1931, is formally definable. Kleene states in a letter to the author of 3 November 1981 that his original October 1933 manuscript also contained the treatment of course-of-values recursion with or without a parameter, this work having been completed before Péter's 1934 paper appeared in print.[24] Kleene also formally defines the least number operator, as we shall see in the next section.

By the time that Kleene came to revise his 1935 paper, in the spring of 1934, be had been attending a course of lectures in Princeton given by Gödel and some of the information that he gained from these lectures prompted him to add a few extra points to his paper. In his lectures Gödel added the identity functions to the constant functions and the successor function to give three sets of initial functions. This addition made the treatment of recursive definitions much neater and easier and removed the necessity of considering possible reductions in the number of variables on the right hand side of the definitions of substitution and recursion.

Previously, in the October 1933 manuscript, Kleene had made no mention of Gödel's 1931 paper, with its class of primitive recursive functions, but now he wished to do so and consequently he added a comment about the formal definability of the constant and identity functions to his footnote on page 219, and then on page 223 of [**85**] he makes the following comment about the formal definability of the whole class of primitive recursive functions:[25]

> It is clear from the foregoing that every function recursive in
> the limited sense of Gödel [**51**] is definable, if we use
>
> $$\lambda fx.f(x), S(\lambda fx.f(x)), S(S(\lambda fx.f(x))), \ldots$$

---

[24]Péter considers course-of-values recursion in some detail in Section 1 of this paper.
[25]See [**85**, p.223].

as formulas for the numbers $1, 2, \ldots$ respectively (thus going over from our theory of positive integers to a like theory of natural numbers) or if we replace natural numbers by positive integers in Gödel's theory.

Kleene then adds the remark that this will allow him to use Theorems I to IV of Gödel's 1931 paper to prove that functions like quotient, remainder, highest common factor, n^{th} prime number and so on are formally definable. Of course Kleene's definition of the least number operator gave him an alternate, more powerful, method for such proofs which might have been easier to use than Gödel's results.

Kleene also adds the following sentence:[26]

It is also true that functions recursive in various more general senses may be defined formally.

and then, in a footnote, he considers an example of double recursion since he proves that:[27]

Given formulas **F** and **G** having the same free symbols, [it is possible] to obtain a formula **H** such that

$$\mathbf{H}(\mathbf{1}, \mathbf{n}) \text{ conv } \mathbf{F}(\mathbf{n}),$$

$$\mathbf{H}(\mathbf{m} + \mathbf{1}, \mathbf{1}) \text{ conv } \mathbf{G}(\mathbf{m}) \text{ and}$$

$$\mathbf{H}(\mathbf{m} + \mathbf{1}, \mathbf{n} + \mathbf{1}) \text{ conv } \mathbf{H}(\mathbf{m}, \mathbf{H}(\mathbf{m} + \mathbf{1}, \mathbf{n})))$$

for $m, n = 1, 2, \ldots$

---

[26]See [**85**, p. 224].
[27]*Ibid.*

This function is actually Ackermann's function, given in the same form apart from the slight change concerning natural numbers as against positive integers, that Gödel used when he introduced it in his 1934 lectures. Kleene states in a letter to the author of 3 November 1981 that, in fact, he first learned of Ackermann's function from Gödel's lectures. Gödel, of course, introduced Ackermann's function into his lectures to illustrate that other, more complicated, forms of recursion existed apart from his 1931 definition and he then proceeded to introduce the concept of general recursive functions which will be explored in the next chapter.

4.6. Decision Problems

The second section of the latter part of Kleene's paper consists of the derivation of some connections between the normal form of formulae in Church's λ-notation and certain decision problems. A decision problem is a problem concerning the existence of a mechanical procedure that is completely described in advance and which can decide the truth or falsity of any one of an infinite class of statements in a finite number of steps.

The particular decision problems considered by Kleene are:

i) the problem of whether solutions in positive integers exist for certain systems of equations connecting functions of n positive integers and, if they do, to actually enumerate these solutions,

ii) similar problems involving functions of n integers or n rational numbers, and

iii) the problem of whether a given formula T in some system of formal logic is provable within that logic or not.

Kleene begins this process by illustrating how it is possible to enumerate all n-tuples of positive integers with some formally definable function. That is, he produces a formally definable function E, say, such that $\mathbf{E}(\imath)$ is the $\imath^{th}$ n-tuple when $\imath$ is the formula for the integer $\imath$.

He then considers the problem of finding solutions in n-tuples of equations such as

$$F(x_1, \ldots, x_n) = G(x_1, \ldots, x_n)$$

$F(x_1, \ldots, x_n)$ and $G(x_1, \ldots, x_n)$ are given functions defined for all n-tuples of positive integers and whose values are positive integers. F and G are formally definable by formulae $\mathbf{F}$ and $\mathbf{G}$.

By using the formally definable function δ_y^x which has the property that

$$\delta_y^x \text{ conv } \mathbf{2} \text{ if } x = y$$
$$\delta_y^x \text{ conv } \mathbf{1} \text{ if } x \neq y$$

or some other similar type of function, he is able to reduce the problems of solving

$$F(x_1, \ldots, x_n) = G(x_1, \ldots, x_n)$$

to problems as to whether or not

$$\mathbf{R}(\imath) \text{ conv } \mathbf{2},$$

where $\mathbf{R}(\imath)$ is a formally definable formula. In the case given above, using δ_y^x, then

$$\delta_{\mathbf{G}(\mathbf{E}(\imath)))}^{\mathbf{F}(\mathbf{E}(\imath)))}$$

If several solutions exist to the equation

$$F(x_1, \ldots, x_n) = G(x_1, \ldots, x_n)$$

then Kleene wishes to enumerate them; for this purpose he introduces his least number operator $\mathfrak{p}$.

This $\mathfrak{p}$ function has the properties that:

If $\mathbf{D(k)}$ conv $\mathbf{2}$ then $\mathfrak{p}(\mathbf{D(k)})$ conv $\underline{k}$ and

if $\mathbf{D(k)}$ conv $\mathbf{1}$ then $\mathfrak{p}(\mathbf{D(k)})$ conv $\mathfrak{p}(\mathbf{D(k+1)})$

Hence $\mathfrak{p}$ will locate each 2 in the set of values of the function $D(k)$. So if $n_1, n_2, \ldots$ are the first positive integers such that $\mathbf{D(k)}$ conv $\mathbf{2}$, then

$$\mathfrak{p}(\mathbf{D(1)}) \text{ conv } \mathbf{n_1}$$

$$\mathfrak{p}(\mathbf{D(n_1+1)}) \text{ conv } \mathbf{n_2}$$

and so on. This means that $\mathbf{E}(\mathfrak{p}(\mathbf{R(1)}))$ gives the first solution to the equation

$$F(x_1, \ldots, x_n) = G(x_1, \ldots, x_n)$$

if one exists, and then $\mathbf{E}(\mathfrak{p}(\mathbf{R(n_1+1)}))$ gives the second solution if the first is given by $\mathbf{E(n_1)}$ and so on.

Finally Kleene defines a function $\mathfrak{P}$ in terms of $\mathfrak{p}$ so that:

$\mathbf{E}(\mathfrak{P}(\mathbf{R(1)}))$ gives the first solution to $F = G$

$\mathbf{E}(\mathfrak{P}(\mathbf{R(2)}))$ gives the second solution to $F = G$

and so on.

This least number operator, $\mathfrak{p}$, plays a central role in Kleene's work on λ-definable functions, and also later in his work on recursive functions. As he says in a letter to the author of 28 July 1977, "I could hardly have progressed far in λ-definability without it." He calls it the $\mathfrak{p}$ function for "perpetual motion."

Kleene is now in a position to be able to enumerate solutions to equations similar to

$$F(x_1, \ldots, x_n) = G(x_1, \ldots, x_n)$$

or even systems of equations of that type. But, even more importantly, he now has a criterion for deciding whether more than a few solutions exist, or even whether solutions exist at all.

This is possible because of the definition of the $\mathfrak{p}$ function, which allows Kleene to be able to demonstrate that:[28]

> If $\mathbf{D}(\imath)$ conv $\underline{1}$ for ever positive integer $\imath \geq$ the positive integer k, then $\mathfrak{p}(\mathbf{D}(\mathbf{k}))$ has no normal form.

or in particular:

> If $\mathbf{D}(\imath)$ conv $\mathbf{1}$ for every positive integer then $\mathfrak{p}(\mathbf{D}(\mathbf{1}))$ has no normal form.

In this way Kleene has reduced the problem of the existence of solutions in positive integers of equations such as:

$$x^t + y^t = z^t$$

to the existence of a normal form for some formally definable formula. Any solutions that do exist will be enumerated by this same formula. We can now begin to see the essence of the importance of normal form.

Kleene proves the result of the non-existence of a normal form for $\mathfrak{p}(\mathbf{D}(\mathbf{k}))$ by considering reductions and quoting two results given in a further paper written by Church and Rosser, "Some Properties of Conversion"[28].

A reduction is a derivation of **B** from **A** using only Rules of Conversion I and II with at least one use of Rule II. That is, a reduction makes a substitution or effectively attempts to evaluate a function of positive integers. Reductions are needed because, by the definition of normal form, all uses of Rule II must have been carried out already. So if a normal form exists it has to be produced by a reduction.

The results needed from Church and Rosser's paper are that:

---

[28]See [**85**, p. 231].

  i) If **P** conv **Q**, then there exists a conversion of **P** into **Q** in which all
     applications of Rule III come after all applications of Rule II.
 ii) If **A** has a normal form then every sequence **A** reduces to **A'** reduces
     to **A''** ... of reductions is finite.

By these means Kleene has produced his first result on decision problems,
that is the reduction of problems involving solutions of equations in positive
integers to problems as to the existence of a normal form of a formally definable
formula. He then indicates how this result can be extended to solutions of
equations in all integers or in all rational numbers. He does this by showing how
to construct positive and negative integers and rational numbers in Church's
λ-notation.

Kleene then continues with his analysis of decision problems by extending
the work done on the decision problem for the equations

$$F(x_1, \ldots, x_n) = G(x_1, \ldots, x_n)$$

to the key decision problem for a system of logic: "Given any formula **T** in the
notation of that system of logic, is **T** provable in that system?"

To deal with this situation Kleene first demonstrates that it is possible to
enumerate with repetitions the formulæ derivable from a given set of the axiom
formulæ $\mathbf{A}_1, \ldots \mathbf{A}_l$ by zero or more operations of a set of formulæ representing
the rules of inference. He then proceeds to borrow Gödel's device of setting up
a 1:1 correspondence between certain positive integers and a class of formulæ
M so that he can use a function of positive integers to enumerate the numbers
corresponding to provable formulæ. That is, he proves the existence of a formula
L such that:

$$\mathbf{L(n)} \text{ conv } \mathbf{m}$$

where m represents the number associated with the nth provable formula.
Kleene can then express the problem of whether T is a provable formula in
a system of logic, as follows:

Does there exist an n such that $\mathbf{L(n)}$ conv $\mathbf{t}$?

where $\mathbf{t}$ is the number associated with formula T. As before he can then use the function

$$\delta_{\mathbf{t}}^{\mathbf{L(n)}}$$

to change this problem to the problem of whether or not, for some formula $\mathbf{D}$, $\mathbf{D(n)}$ conv $\mathbf{2}$.

Then he utilizes the $\mathfrak{p}$ function to find the first n such that $\mathbf{D(n)}$ conv $\mathbf{2}$ if such an n exists, exactly as in the previous decision problem. This means, as before, that if $\mathbf{2}$ does not exist, then some formula does not have a normal form. Finally, using the $\mathfrak{P}$ function, Kleene is able to claim:[29]

> The problem of whether $\mathbf{T}$ is provable in the system of logic
> is equivalent to the problem of whether some stated formula
> [actually the formula is $\mathfrak{P}(\lambda\mathbf{n}.\, \delta_{\mathbf{t}}^{\mathbf{L(n)}}, 1)$] has a normal form or
> not.

This analysis will work for any system of formal logic F, for which the condition is satisfied that there is a class M of formulæ such that

   i) all provable formulæ of F belong to M,
  ii) T belongs to M,
 iii) there exists a one to one correspondence of M to a class of positive
      integers such that the numbers corresponding to provable formulæ can
      be enumerated formally.

In particular the system of *Principia Mathematica* satisfies this condition. Kleene has thus shown that the problem of T being provable in the system of

---

[29]Adapted from [**85**, p. 233].

Principia Mathematica is dependent on the existence of a normal form for some
formally definable formula.

4.7. Conclusion - The Inconsistency of Church' s Logic

The third and final section of Kleene's paper consists of results needed by
Kleene and Rosser to prove Church's system of logic inconsistent, all of this
section being added to the paper when Kleene revised the paper. This section
is therefore best considered in conjunction with the paper, "The Inconsistency
of Certain Formal Logics" [**96**] where Kleene and Rosser actually produce this
proof of the inconsistency of Church's system of logic. Their joint paper also
contains a proof of the inconsistency of the combinatorial logic of Curry and
other similar logics.

The major part of the last section of Kleene's revised paper is devoted to
establishing a representation of Church's logic within itself in the manner of
Gödel in 1931. In practice only a few of Church's original set of postulates had
ever been used; namely axioms 1, 3 through 11, and 14 through 16 together
with rules of procedure I through V. Thus Kleene considers just that part of
the logic and designates it C_1.

He achieves his, representation of C_1 within C_1 by utilizing some work of
Rosser from "A Mathematical Logic Without Variables" [**136**] which connects
formulæ of C_1 without free symbols with certain combinations without free
symbols in Curry's system of logic. He then proves that these combinations can
be converted into a special type of combination, called a uniform combination,
which can easily be coded into formal lists of 1's and 2's. He finally shows
that these lists of 1's and 2's can be correlated with certain formulæ within the
system C_1 called metads.

These metads therefore represent combinations and, in turn, these combinations represent formulae in C_1 It can be shown that these metads, preserve the properties of the original formulæ in that it is possible to prove that functions of these metads, that correspond to metamathematical notions relating to the form of the formulae, can be defined within the logic. For example, if the metad **a** represents the formula **F(Q)** then a function exists to produce the metad $\mathbf{a}_1$ that represents **F** and the metad $\mathbf{a}_2$ that represents **Q** and vice versa.

At this stage Kleene introduces certain formulæ which produce important enumerations. First he succeeds in defining a formula **H** that will enumerate with repetitions all the metads that represent provable formulæ in C_1. Secondly he uses this **H** function to produce a formula **U** with the property that, if **F(U)** is provable in C_1 and **Q** contains no free symbols, then there exists a positive integer q such that **U(q)** conv **Q**. That is, the formula **U** enumerates formulæ **Q** from provable functions of the form **F(Q)**. In the 1935 paper Kleene does not specify the form of the function **F** but needs the assumption that there exists some formula **P** such that **F(P)** is actually provable, so that there is at least one formula in the enumeration.

We can now turn to the joint paper by Kleene and Rosser, "The Inconsistency of Certain Formal Logics," which was also published in 1935. In this paper the function used for **F** is given as

$$\mathbf{F} \implies \lambda f[N(x) \supset_x N(f(x))]$$

which can be interpreted to mean "if x is a positive integer then the function f is a single valued function of a positive integer x which takes integer values".

So if we can prove**F(P)** then P must be a function with positive integers as arguments and with integral values. Hence we can see that the formula **U** is enumerating formulæ for which it is possible to prove the formal theorem that they represent functions of positive integers which take integer values.

The final theorem giving the inconsistency of Church's system of logic is then derived by proving that $\lambda n \mathbf{U}(n,n)$ and hence $\lambda n[\mathbf{1} + \mathbf{U}(n,n)]$ are formulæ that represent functions of a positive integer with integer values. That is,

$$\mathbf{F}(\lambda n[\mathbf{1} + \mathbf{U}(n,n)])$$

is provable formally, and hence there exists a positive integer q such that:

$$\mathbf{U}(\mathbf{q}) \text{ conv } \lambda n[\mathbf{1} + \mathbf{U}(n,n)].$$

From this we obtain:

(4.1) $$\mathbf{U}(\mathbf{q},\mathbf{q}) \text{ conv } \{\lambda n[\mathbf{1} + \mathbf{U}(n,n)]\}.(\mathbf{q})$$

(4.2) $$\text{conv } \mathbf{1} + \mathbf{U}(\mathbf{q},\mathbf{q})$$

and then it is a simple step to prove that:

(4.3) $$\mathbf{1} \text{ conv } \mathbf{2}.$$

Therefore the system C_1 is seen to be inconsistent.

Rosser had, in fact, produced an outline of his plan for proving the inconsistency of Church's system in the autumn of 1933, and had immediately informed Church. In December 1933 or January 1934 Rosser wrote to Kleene informing him of this plan and succeeded in recruiting him to the task of proving this inconsistency. Kleene had left Princeton in June 1933 and he recalls in a letter to the author of 3 November 1981 that he started work on this proof in his farmhouse in Maine before he was offered a Research Assistantship to continue the work at Princeton, where he would be able to work more efficiently. Kleene accepted and returned to Princeton on 7 February 1934. By about the end of March or April they were pretty sure that the actual proof of inconsistency could be completed.

One of the effects of Rosser's discovery of a possible inconsistency in Church's system was that Kleene took back his two papers, "Proof by Cases in Formal Logic" [84] and "A Theory of Positive Integers in Formal Logic" [85] and revised them, the revised papers being resubmitted on 28 March 1934 and

18 June 1934 respectively. The major part of this revision was the addition of Section 19 to [**85**], which contained the representation of C_1 within itself and the formulæ **U** and **F** and so forth, these notions being added specifically for their use in the subsequent proof of the inconsistency. In his 1935 paper Kleene did not need to specify a particular function for **F** since he was able to prove a more general result.

Kleene states in his letter to the author of 13 November 1981 that the idea for representing C_1 in C_1 of course owed something to Gödel but that:

> My particular method of representation., using the metads, is different from Gödel's method of numbering. If my recollection is correct, I started out with a Gödel-like numbering, and then decided that in my context the metads were available and easier to use.

Another effect of Rosser's discovery of a possible inconsistency was that Church started to propose the possibility of extracting out the notion of λ-definability from the formal logic. Church also responded by writing and delivering his address entitled, "The Richard Paradox" at the meeting of the Mathematical Association in Cambridge, Massachusetts. on 30 December 1933. As Kleene says in his 3 November 1981 letter to the author:

> Rosser told me very explicitly that writing "The Richard Paradox" was a response by Church to his [Rosser's] proposal of his plan to derive a contradiction in Church's system.

In this address Church analyzes the relationship between the numbers of possible formulæ in any system of logic similar to his own and the number of functions of a positive integer. He quotes the function **F** and talks about its properties. He also shows knowledge of the formula **U** by stating that a method exists within the logic to enumerate all the formulae for which it is possible to prove the theorem that they represent functions of a positive integer. Hence there is an enumeration

$$\mathbf{Q}_1(\mathbf{x}), \mathbf{Q}_2(\mathbf{x}), \ldots$$

of formulae for which the proof that they represent functions of a positive integer can be carried out formally within the system. But he also shows, by utilizing the familiar diagonal argument, that the number of possible functions of a positive integer is non-enumerable. He therefore concludes that any such system of logic must be inadequate to represent arithmetic.

Due to his loss of faith in his system of logic, Church began to formulate a new system of logic that would avoid his awn proof of inadequacy and also Kleene and Rosser's proof of inconsistency. He analyzed why the proofs could be carried out in his system of logic and decided that they relied on the fact that it was possible to recognize a formula, of the form:

$$\lambda f[N(x) \supset_x N(f(x))]$$

So Church set about designing a new system with a non-enumerable multiplicity of definitions of implication, for which there was no uniform means of determining whether or not a given formula was an implication symbol. Thus the formulae $\mathbf{Q}_i$ could not be enumerated effectively and hence both his proof of inadequacy and Kleene and Rosser's proof of inconsistency would no longer apply to his new system.

He disclosed the beginning of the existence of this new system in his 1933 address where he said:[30]

> As I speak, I have in mind a particular set of postulates for symbolic logic, whose freedom from contradiction can be proved, and which lead to a non-enumerable multiplicity of definitions of implication, in the manner we desire.

---

[30]See [**19**, p. 360].

This new system of Church's was rather impractical and has not subsequently found favor.

Once the predecessor function had been proven to be formally definable, Church began to speculate that perhaps all effectively calculable functions were formally definable. He first asserted the thesis, that these two classes of functions were equivalent, to Kleene in February or March of 1934.

A little while later Gödel introduced the concept of general recursive functions and it was conjectured that perhaps these were also equivalent to the class of effectively calculable functions. Part of the ultimate justification of the thesis equating these three classes of functions, later called Church's thesis, was the proof that the class of general recursive functions and the class of formally definable functions were equivalent. In the next chapter we shall consider the development of the general recursive functions.

CHAPTER 5

General Recursive Functions

5.1. Introduction

In this chapter we shall see the theory of recursive functions finally emerge as a fully fledged theory of mathematics. Gödel's 1931 paper gave the main impetus for the study of recursive functions and this chapter will be concerned with the development in the years immediately following.

In 1931 Gödel gave the first precise definition of the class of ordinary recursive functions. In some of the previous papers of the Hilbert school various examples had been given of functions that were not recursive in the sense of Gödel's definition but were recursive in a more general sense. Of these functions Ackermann's function was probably the most well known. During the next four or five years the development of recursive functions proceeded in two directions. First there was the continued analysis of the ordinary recursive functions of Skolem and Gödel and second there was the expansion of the concept into a more general definition of a recursive function.

The first part of this chapter will consider the investigations of researchers like Gödel, Hilbert and Bernays and, above all, Rózsa Péter into the boundaries of what constituted an ordinary recursive function. Péter, in her 1934 paper, referred to the ordinary recursive operation as primitive recursion and Kleene, in his 1936 paper [86], introduced the term "primitive recursive" function for this type of function. Since then this name has become standard. All this research led to the definition of these primitive recursive functions being changed in various ways, such as increasing or decreasing the number of initial functions, changing the number of parameters, allowing several functions to be defined at once or permitting the nesting of functions inside themselves. In each case it was examined whether or not the changes produced functions that exceeded the class of primitive recursive functions. By the end of this period the concept of primitive recursive functions bad been made much more definite and other methods of producing non-primitive recursive functions were thereby discovered.

The second part of this chapter is concerned with the development of a more general concept of a recursive function. Once primitive recursive functions and arithmetic were introduced in the 1920's they were rapidly accepted as constituting the finitary methods required by Hilbert. The study of functions that could be calculated by a finite process, like these primitive recursive functions, was given an impetus by various of the papers produced by the Hilbert school. The demand for a more general concept of the class of effectively calculable functions arose from the need to solve various decision problems within logic or some other formal framework.

Obviously the possibility of solving, or proving unsolvable, general decision problems, like the determination of some finite procedure for deciding whether or not a certain formula is derivable from the axioms of a logic, depends on the criteria of what constitutes a finite procedure. By the use of Gödel numbering the finite procedure can be changed into a number-theoretic function and the criteria become whether or not the functions involved are effectively calculable. Now all primitive recursive functions are effectively calculable, but so are other more general functions formed by extending the primitive recursive functions such as Ackermann's function or by diagonalizing the whole class. So by the late 1920's and early 1930's there was a need for a more general concept of an effectively calculable function and the first real breakthrough towards this objective came about through the work of Herbrand and Gödel.

We will investigate the development of the Herband-Gödel formulation of the class of general recursive functions from Herbrand's original suggestion to Gödel's much improved version. We will also consider Kleene's formal definition of this class of functions contained in [**86**]. In this paper Kleene gave his very important normal form theorem for these functions and also produced the first results in recursive function theory proper.

All of this work gains in significance when Church's thesis, which equates the class of general recursive functions with the class of effectively calculable functions, is taken into account. The later papers of this era need to be read with Church's thesis in mind and in the next chapter more details and evidence for Church's thesis will be considered.

5.2. The Development of Ordinary Recursive Functions

By his significant use of recursive functions in his 1931 paper Gödel firmly established their place in foundational research. He gave the first precise definition ar the class of functions involved and gave the now standard method of introducing recursive relations via representing functions. He proved several theorems on recursive relations including the method of eliminating bounded quantifiers. He made significant use of them as a metamathematical tool and indicated their use as a part of a constructive function defining apparatus. He also gave a definition of the class of decidable relations and proved that it contained the class of recursive relations. This showed that every recursive relation was decidable. The topic of decidability is discussed in more detail in the next chapter.

In both their 1934 and 1939 volumes of *Grundlagen der Mathematik* Hilbert and Bernays, use recursive functions. In their 1939 volume they formulated all the details of Gödel's work by giving the full proofs of the theorems where Gödel had only indicated a proof. In addition, they amplified Gödel's results for different formal systems. Their only major new contribution to the theory of recursive functions was to prove that a schema for the introduction by recursion of several functions at once still gave functions that were ordinary recursive functions.

The first person to do any substantial research into ordinary recursive functions after 1931 was Rózsa Péter who, in an address given at Zürich in 1932, announced that, subsequent to their significant use as aids to research by Hilbert and Ackermann and by Gödel, she intended to research these functions for their own sake.

In two papers that belong to the era prior to Gödel's lectures at Princeton ([119],[120]), she produced some of the many results she was to contribute to the theory of recursive functions. These papers belong to the era prior to Gödel lectures in the spring of 1934 in Princeton because some of the anticipated results were already announced at Zürich in 1932 and, since the papers were submitted for publication in April 1934 and July 1934 respectively, they had obviously been researched before Gödel's lectures. In these lectures Gödel made some minor changes to the definition of recursive functions and also initiated research into a much wider class of functions, the general recursive functions. His lectures thus represent a watershed in recursive function theory.

On page 613 of the 1934 paper Péter uses the term "primitive recursion" to refer to the ordinary recursive operation of Gödel and Skolem and, as mentioned already, after Kleene's use of the term "primitive recursive functions" this terminology became standard. The new results obtained by Péter can be outlined as follows.

First, Péter proves that the definitions of ordinary recursive functions given by Gödel in 1931 and by Hilbert in the 1920's are equivalent. In the 1920's Hilbert gave a definition of ordinary recursive functions that did not include higher types of functions as variables. Namely:

$$\phi(0, a_1, \ldots, a_k) = \alpha(a_1, \ldots, a_k)$$

$$\phi(n + 1, a_1, \ldots, a_k) = f_{b_1, \ldots, b_k}(n, a_1, \ldots, a_k, \phi(n, b_1, \ldots, b_k))$$

In this definition substitutions have been made for the parameters, that is the $b_1, \ldots, b_k$ could represent previously defined recursive functions. Péter proves that this definition gives the same class of functions as does Gödel's definition of 1931, that is, both give primitive recursive functions.

Second, Péter proves that "course-of-values recursion" can be reduced to primitive recursion.

Course-of-values recursion defines a function for argument $y + 1$ in terms of one or more of the values of the function for argument s, where $s \leq y$. Skolem, in 1923, introduced the idea of course-of-values recursion and proved its equivalence to his ordinary recursion. But Skolem only considered a special case where the function was defined in terms of one other value of the function for argument $s \leq y$ rather than the general case considered by Péter. Also Skolem's proof consisted of an informal verbal justification in terms of the constructive nature of the new definition.

Third, Péter proves that nested simple recursion reduces to primitive recursion.

The term "simple recursion" refers to recursion on one variable only and examples of functions defined by nested simple recursion are:

$$\phi(0, a) = \alpha(a)$$

$$\phi(n + 1, a) = \beta(n, \phi(n, \gamma(n, a, \phi(n, a))))$$

and

$$\phi(0, a) = \alpha(a)$$
$$\phi(n + 1, a) = \phi(n, \phi(n, t))$$

where t is a function of a and n.

Notice that the function ϕ occurs in one or more of the argument places of ϕ itself in the right hand side of these definitions.

Péter proves that these functions are primitive recursive but in contrast it was already known that nested double recursion led out of the class of primitive recursive functions, namely Ackermann's function from 1928 [2]. In [121] functions defined by double recursion without nesting are proved to be reducible to primitive recursive functions. Recall that Skolem had already used a double recursive function of this kind in 1923.

In summary, double recursion on its own or nested recursion on its own does not lead out of the class of primitive recursive functions, but nested and double recursion together do lead out of the class.

Fourth, Péter shows that it is possible to produce the same class of primitive recursive functions by reducing the number of parameters in the schema for definition by recursion.

That is, the definition

$$\phi(0, a) = \alpha(a)$$
$$\phi(n + 1, a) = \beta(n, a, \phi(n, a))$$

can be replaced by the schema

$$\psi(0) = 1$$
$$\psi(n + 1) = \psi(n, \psi(n))$$

and in principle a reduction can be made from

$$\phi(0, a_1, \ldots, a_k) = \alpha(a_1, \ldots, a_k)$$
$$\phi(n+1, a_1, \ldots, a_k) = \beta(n, a_1, \ldots, a_k, \phi(n, a_1, \ldots, a_k))$$

as well. This reduction is made at the cost of adjoining new initial functions. Subsequently both Péter and R. Robinson in [133] reduce even further the schema necessary for defining primitive recursive functions.

Fifth, Péter produces a better proof that Ackermann's function is not primitive recursive and also produces another non-primitive recursive function by directly utilizing the diagonal method.

Hence in these and later papers Péter made substantial improvements to the theory of primitive recursive functions by analyzing several different modes of definition for recursive functions, most of these being already in existence, to determine whether or not they defined the class of primitive recursive functions. The fruits of all this work can be found in [122].

The last substantive improvement to the theory of primitive recursive functions in the early 1930's was the addition of a third set of initial functions to Gödel's list in his definition of 1931. There Gödel had given the successor function, $x + 1$, and the constant functions, $f(x_1, \ldots, x_n) = c$ as his initial functions, but by 1934 Gödel realized that including the identity functions, $U_j^n(x_1, \ldots, x_n) = x_j, (1 \le j \le n)$, in his list would simplify and improve the definition of the primitive recursive functions.

Gödel's [54] is a paper produced from his lectures at Princeton in the spring of 1934. At the suggestion of Oswald Veblen, Rosser and Kleene produced a set of notes from Gödel's lectures and these notes were mimeographed and distributed at the time. They have since been published [35]. Gödel's lectures represented a reworking and updating of his 1931 paper and several new results and changes of emphasis can be seen in his 1934 paper.

First, the equivalent result to Theorem V of 1931 and not numbered in 1934 is expressed more clearly in that Gödel gives a more detailed description of what he means by the representation of functions by formulæ of his formal system. That is:

> If z_0 is an abbreviation for the formal expression 0, if z_1 is an abbreviation for the formal expression $N(0)$ and so forth where N is the successor function, then the z_i's will represent the natural numbers in the formal logic. Then if $\phi(x_1, x_2, \ldots)$ is a function of positive integers we have that:[1]
> $\mathbf{G}(\mathbf{u_1}, \mathbf{u_2}, \ldots)$ *represents* $\phi(\mathbf{x_1}, \mathbf{x_2}, \ldots)$ if
>
> $$\mathbf{G}(\mathbf{z_m}, \mathbf{z_n}, \ldots) = \mathbf{z}_{\phi(m, n, \ldots)}$$
>
> is provable formally for each given set of the natural numbers $m, n, \ldots$; in other words if $\mathbf{G}(\mathbf{z_m}, \mathbf{z_n}, \ldots) = \mathbf{z}_k$ is provable formally whenever $\phi(m, n, \ldots) = k$ holds.

He gives a similar definition for relations:[2]

> If $\mathbf{R}(\mathbf{x_1}, \mathbf{x_2}, \ldots)$ is a class or relation of natural numbers, we shall say that the formal functional expression $\mathbf{H}(\mathbf{u_1}, \mathbf{u_2}, \ldots)$ *represents* $\mathbf{R}(\mathbf{x_1}, \mathbf{x_2}, \ldots)$ if we can prove formally $\mathbf{H}(\mathbf{z_m}, \mathbf{z_n}, \ldots)$ whenever $\mathbf{R}(\mathbf{m}, \mathbf{n}, \ldots)$ holds, and $\sim \mathbf{H}(\mathbf{z_m}, \mathbf{z_n}, \ldots)$ whenever $\mathbf{R}(\mathbf{m}, \mathbf{n}, \ldots)$ does not hold.

As we shall see later, these results are essentially equivalent to Gödel's definition of a function being "reckonable" in a system. Gödel then proceeds to prove that all primitive recursive functions, classes and relations are representable by formulae in the formal system. In 1931 Gödel only indicates a

---

[1]See [**35**, p. 58].
[2]Ibid.

method of producing a proof whereas in 1934 Gödel can give the proof easily since his formal system, now permits variables for functions.

Second, Gödel's proof of his main result from 1931, Theorem VI, is also much shorter and simpler and follows the lines of the informal account given by Herbrand [**67**].

Third, Gödel includes a discussion of the relationship between the paradoxes and the arguments used in his paper and in conjunction with this he discusses the possibility of defining truth of a language within that language. This work has been analyzed more fully in papers by Tarski [**159**] and Carnap [**16**].

Finally, perhaps the most important updating of Gödel's 1931 paper occurs in his discussion of a formal system and the accompanying emphasis on constructive methods.

In his 1931 paper Gödel names two formal systems, the system of *Principia Mathematica* and the Zermelo-Fraenkel system for set theory. He discusses briefly some of the properties these systems share, namely:[3]

> [the] proofs can be carried out according to a few mechanical rules.

and

> Both of these systems are so broad that all methods of proof used in mathematics today can be formalized in them.

He then uses essentially the system of *Principia Mathematica*.

_____

[3]See [**35**, p. 5].

In 1934 Gödel describes a formal system only slightly different from the one used in 1931, but he also discusses in detail conditions that a formal system would have to satisfy in order that the undecidable results he has produced should apply. Most of these conditions are couched in terms of constructivity and recursive functions. For instance, having defined terms, such as rules of inference, meaningful formulæ, and axiom and so forth in his formal mathematical system, he makes an important restriction on their use:[4]

> We require that the rules of inference, and the definitions of meaningful formulæ and axioms, be constructive; that is, for each rule of inference there shall be a finite procedure for determining whether a given formula **B** is an immediate consequence (by that rule) of given formulas $\mathbf{A}_1, \ldots, \mathbf{A}_n$, and there shall be a finite procedure for determining whether a given formula **A** is a meaningful formula or an axiom.

Later in the paper he states that a precise condition that the class of axioms and the relation of immediate consequence should be constructive is that if the symbols and formulæ are Gödel numbered then the class of axioms and the relation of immediate consequence should be primitive recursive.

Other conditions on the general formal system include, the relationship between the natural number n and the number representing z_n should be primitive recursive, and that all primitive recursive relations should be representable in the system. Also the system must contain symbols for negations, and "for all x" or something essentially equivalent.

Finally the system needs to be consistent and ω-consistent in the sense that:

---

[4]See [**35**, p. 41].

Consistency: If $\mathbf{R}(\mathbf{v}, \mathbf{w})$ is a primitive recursive propositional function of two variables, then

$$\mathbf{R}(\mathbf{z}_p, \mathbf{z}_q) \text{ and } \sim \mathbf{R}(\mathbf{z}_p, \mathbf{z}_q)$$

shall not both be provable.

ω-*Consistency*: If $\mathbf{F}(\mathbf{v})$ is a primitive recursive propositional function of one variable, then the formulæ

$$\sim (\forall \mathbf{v})\mathbf{F}(\mathbf{v}), \mathbf{F}(\mathbf{z}_0), \mathbf{F}(\mathbf{z}_1), \ldots$$

shall not all be provable.

He does not list the further conditions under which it would be possible to produce a formal proof of his Theorem XI about a system containing a proof of its own freedom from contradiction but states that:[5]

> ... they are conditions satisfied by all systems of the type under consideration which contain a certain amount of ordinary arithmetic.

All of these conditions show the importance with which he regards recursive procedures in describing a constructively defined formal system. This is amplified by his statements on general recursive functions given in the next section.

5.3. General Recursive Functions

During the course of his 1934 lectures Gödel says:[6]

---

[5]See [**35**, p. 63].

[6]See [**35**, p. 43–44].

Recursive functions have the important property that, for each given set of values of the arguments, the value of the function can be computed by a finite procedure. Similarly, recursive relations (classes) are decidable in the sense that, for each given n-tuple of natural numbers, it can be determined by a finite procedure whether the relation holds or does not hold (the number belongs to the class or not), since the representing function is computable.

Significantly, he adds in a footnote:

The converse seems to be true, if, besides recursions according to the scheme (2), [primitive recursions], recursions of other forms; (e.g. with respect to two variables simultaneously) are admitted. This cannot be proved, since the notion of finite computation is not defined, but it serves as a heuristic principle.

This footnote suggests that all functions computable by a finite procedure are recursive, providing we allow other definitions of recursion. Since later in the paper Gödel defines general recursive functions, this looks very much like a statement of Church's thesis, which equates functions computable by a finite procedure with the class of general recursive functions, but predating Church's 1936 paper. However, Gödel states in a letter to Davis[7] that he was not sure at the time of these lectures that his general recursive functions comprised all possible recursions, so that his footnote is not a statement of Church's thesis.

Actually it is irrelevant whether this is or is not a statement of Church's thesis since Church conceived of his thesis before Gödel came to Princeton and introduced his notion of general recursive functions. Church had already been

---

[7]See [**35**, p.40].

speculating on whether or not the class of λ-definable functions embraced all effectively calculable functions for some time and had, in fact, asserted the thesis to Kleene early in 1934. The evidence for this comes from several sources.

i) *Church's paper*:[8] The question of the relationship between effective calculability and recursiveness (which it is here proposed to answer by identifying the two notions) was raised by Gödel in conversation with Church. The corresponding question of the relationship between effective calculability and λ-definability had previously been proposed by the author independently.

ii) *Letter to the Author from Kleene, 9 September 1982*: "I can date the time Church definitely asserted his thesis to me as in February or March 1934, as I only returned to Princeton (having been absent during the fall of 1933) on February 7, 1934. It was later, in April or May 1934, that Gödel came out with his notion of general recursiveness."

iii) *Letter from Church to Kleene, 9 November 1935*: "In regard to Gödel and the notions of recursiveness and effective calculability, the history is the following. In discussion with him the notion of λ-definability, it developed that there was no good definition of effective calculability. My proposal that λ-definability be taken as a definition of it he regarded as thoroughly unsatisfactory. I replied that if he would propose any definition of effective calculability which seemed even partially satisfactory I would undertake to prove that it was included in λ-definability. His only idea at the time was that it might be possible, in terms of effective calculability as an undefined notion, to state a set of axioms which would embody the generally

---

[8]See [**23**, p. 356, footnote 18].

accepted properties of this notion, and to do something on that basis. Evidently it occurred to him later that Herbrand's definition of recursiveness, which had no regard to effective calculability, could be modified in the direction of effective calculability, and he made this proposal in his lectures. At that time he did specifically raise the question of the connection between recursiveness in this new sense and effective calculability, but said that he did not think that the two ideas could be satisfactorily identified "except heuristically".[9]

These show that the credit for the original formulation of Church's thesis must undoubtedly go to Church. The last quotation also illustrates quite clearly the history behind the development of general recursive functions in Gödel's lectures. Indeed these discussions between Church and Gödel obviously account for the addition of footnote number 3 to Gödel's paper, as the last part of this letter indicates; but, as already said, Gödel was sceptical about the thesis and remained so until after Turing's work [160]. Thus, Church was justifiably the first in print with the thesis and its evidence in [23].

The notion of general recursive functions, then, was introduced by Gödel in his lectures at Princeton in 1934. These functions are more general than, and include, all the primitive recursive functions. His idea for these functions derived from some results produced by Herbrand. Consequently we first consider Herbrand's generalization of the notion of primitive recursive functions before discussing Gödel's definitive definition for the class of general recursive functions.[10]

---

[9]This view is substantiated by statements in print such as in Crossley's *Reminiscences of Logicians* [29, p. 6] and in Kleene's "The Work of Kurt Gödel" [93, p. 769].

[10]It should be noted that Church, in a letter to the author of 21 April 1978, expressed the opinion that the notion of general recursive function should nevertheless be credited to Gödel since Herbrand's formulation was far from adequate.

Primitive recursive functions are formed from certain initial functions by substitution and the recursive schema:

$$\phi(0, x_2, \ldots, x_k) = \psi(x_2, \ldots, x_k)$$
$$\phi(n+1, x_2, \ldots, x_k) = \mu(n, \phi(x_2, \ldots, x_k), x_2, \ldots, x_k).$$

This schema is essentially equivalent to Skolem's original conception but the formulation is due to Gödel. Skolem, himself, explored the possibility of generalizing the notion of primitive recursive functions by investigating "course-of-values" recursion, but found it did not give a larger class of functions. Hilbert was the first to produce an extension to the class of recursive functions by essentially keeping the general form of the above schema but allowing the use of functions of higher types as variables. Ackermann proved in his 1928 paper that functions did exist that could not be reduced to definitions using the simple schema above and that these functions could be defined without recourse to variables of higher types. He used nested double recursion to define his function. Ackermann defined his function as:

$$\phi(a, b, 0) = a + b$$
$$\phi(a, 0, n+1) = \alpha(a, n)$$
$$\phi(a, b+1, n+1) = \phi(a, \phi(a, b, n+1), n),$$

where $\alpha(a, n)$ is a primitive recursive function previously defined.

Thus, it was known that non-primitive recursive functions existed, but Ackermann's function did not indicate a general method of producing all recursive functions.

The two essential features common to all the recursive functions defined at this time were:

i) the iterative method of definition in terms of an induction on one or more variables

ii) the finite method of determining the value of the function from any given argument value using formal equations.

It was Herbrand who took the bold step of generalizing the nation of recursive function by taking feature (ii) and abandoning the iterative or inductive part of the definition. Herbrand did not set out specifically to generalize the notion of recursive function, so he did not formulate his ideas this way or name them with any name such as "general recursive functions". Nor did he come to this concept in one step.

Herbrand was working on proving arithmetic consistent and thus his ideas were very much related to the Hilbert school. This work was still prior to Gödel's 1931 paper, although it partly overlapped with the publication of Gödel's results. Herbrand, in his 1930 paper [3], was considering intuitionistic arguments[11] and noted that all the functions introduced should be computable for all values of their arguments by methods that were completely described in advance.

Hence, when Herbrand, in his 1932 paper [67], was presenting his axioms for a part of arithmetic which he intended to prove consistent, he included a set of hypotheses which involved general finitary functions. These functions, defined by hypotheses C, were in essence his conception of this class of all the functions uniquely computable by means of operations that were completely described in advance, though he did not give any of the operations he would require, nor did he give a method of checking whether any given function was thus computable. His definition proceeds as follows:[12]

---

[11]At the time of Herbrand's work it was fairly common for members of the Hilbert school to equate intuitionistic with finitary and Herbrand tended to use the term intuitionistic with Hilbert's metamathematics in mind where today we would consider that his meaning was closer to Hilbert's term finitary.

[12]See [64, p. 624].

We can also introduce any number of functions $f_i(x_1, \ldots, x_{n_i})$
together with hypotheses such that

   i) The hypotheses contain no apparent variables;
   ii) Considered intuitionistically, [that is, considered as a
       property of integers and not formally] they make the ac-
       tual computation of the $f_i(x_1, \ldots, x_{n_i})$ possible for every
       given set of numbers, and it is possible to prove intuition-
       istically that we obtain a well-determined result.

He did not name these functions but illustrated the type of function he
had in mind by giving two examples, the first was a function defined by primitive recursion and the second was Ackermann's function although he called it
Hilbert's function.

Herbrand therefore realized his set of functions contained all previous recursive functions but it is not obvious whether, at this stage, he thought his
functions represented all recursive functions. He was, though, well aware of
their great power and generality since he used them to explain why Gödel's
theorem would not apply to his theory.

Although Herbrand researched his 1932 paper prior to Gödel's theorems
being published, he did have time to study Gödel's work and include a final
section in his paper, where he summarized Gödel's theorems and showed why
his part of arithmetic was not bound by Gödel's results. His proof basically consisted of a diagonal argument constructing an intuitionistically defined function
that would not be included in any listing of functions defined by his hypotheses
C.

Gödel had presupposed that all the objects occurring in a proof could be
numbered, and Herbrand's diagonal argument produced an object not in any

possible numbering, thus making Gödel's theorem inapplicable. Herbrand concluded that it was impossible, in any arithmetic containing hypotheses such as C, to formalize Gödel's argument about the arithmetic.

Herbrand had therefore suggested a definition of recursive functions that utilized the finite method of calculation but did not specify an iterative nature to the equations involved.

He obviously considered these functions further because in 1931 he wrote to Gödel and made his ideas much clearer as he suggested the following definition of every recursive function:[13]

> If ϕ denotes an unknown function, and $\psi_1, \ldots, \psi_k$ are known functions, and if the ψ's and the ϕ's are substituted in one another in the most general fashions and certain pairs of the resulting expressions are equated, then if the resulting set of functional equations has one and only one solution for ϕ, ϕ is a recursive function.

It is now clear that Herbrand envisaged sets of equations that could be substituted into each other to calculate values for ϕ. There is no iteration structure to the equations, though the ψ_i, could be previously known primitive recursive functions, and in practice some sort of inductive method would be needed to evaluate ϕ for example using Gödel's definition.

So by this stage Herbrand had realized that his functions represented some sort of generalized version of the primitive recursive functions, whereas in his paper he had not actually stated that he thought his hypotheses C led to a class of functions that could represent every recursive function.

---

[13]See [**35**, p. 70].

Herbrand did not, however, specify any rules of procedure for calculating the values of the function and this has led Wang and others to suggest that Herbrand's functions should be identified with the class of effectively definable functions.[14]

The last stage in producing the functions that are today called general recursive functions was supplied by Gödel. Gödel realized that the computation of all these functions needed to proceed by exactly the same rules, so in his 1934 lectures he made some restrictions on Herbrand's definition and formulated a set of derived equations with well-determined rules of construction. These derived equations allowed the defined function ϕ to be evaluated for any given set of arguments $(n_1, \ldots, n_k)$ in such a way that only previously calculated values of ϕ would be needed at any stage in the computation. That is, some sort of inductive procedure needed to be adopted when ϕ came to be evaluated.

The first restriction Gödel made to Herbrand's definition was:[15]

> ...the left-hand side of each of the given functional equations defining ϕ shall be of the form
>
> $$\phi(\psi_{i_1}(\mathbf{x}_1, \ldots, \mathbf{x}_n), \psi_{i_2}(\mathbf{x}_1, \ldots, \mathbf{x}_n), \ldots, \psi_{i_j}(\mathbf{x}_1, \ldots, \mathbf{x}_n))$$

in other words the left-hand side should only contain one occurrence of the unknown function ϕ.

The second restriction was in terms of derived equations defined:[16]

> 1a) Any expression obtained by replacing all the variables of one of the given equations by natural numbers shall be a derived equation.

---

[14]See [**166**, p. 45].
[15]See [**35**, p. 70–71].
[16]Ibid.

1b) $\psi_{i_j}(\mathbf{k}_1, \ldots, \mathbf{k}_n) = \mathbf{m}$ shall be a derived equation if $\mathbf{k}_1, \ldots, \mathbf{k}_n$ are natural numbers, and $\psi_{i_j}(\mathbf{k}_1, \ldots, \mathbf{k}_n) = \mathbf{m}$ is a true equality

2a) If $\psi_{i_j}(\mathbf{k}_1, \ldots, \mathbf{k}_n) = \mathbf{m}$ is a derived equation, the equality obtained by substituting $\mathbf{m}$ for an occurrence of $\psi_{i_j}(\mathbf{k}_1, \ldots, \mathbf{k}_n) = \mathbf{m}$ in a derived equation shall be a derived

2b) If $\phi(\mathbf{k}_1, \ldots, \mathbf{k}_l) = \mathbf{m}$ is a derived equation where $\mathbf{k}_1, \ldots, \mathbf{k}_l, \mathbf{m}$ are natural numbers, the expression obtained by substituting $\mathbf{m}$ for an occurrence $\phi(\mathbf{k}_1, \ldots, \mathbf{k}_l)$ on the right-hand side of a derived equation shall be a derived equation.

Gödel's second restriction stated:[17]

> ...for each set of natural numbers $\mathbf{k}_1, \ldots, \mathbf{k}_l$ there shall be one and only one m such that $\phi(\mathbf{k}_1, \ldots, \mathbf{k}_l) = \mathbf{m}$ is a derived equation.

All of this is to say that ϕ can only be calculated using derived equations. The construction of the derived equations gave the well-determined rules for evaluating the function ϕ. For any general recursive function defined in terms of known functions $\psi_1, \ldots, \psi_k$ by a set of functional equations the same evaluation procedure applied in each case, namely via the rules of construction of these derived equations.

Thus Gödel had generalized the notion of recursive function by stipulating that the equations giving the values of the function were formally derivable from the equations of recursion by using a substitution rule and a replacement rule and that one, and only one, equation of the form

$$\phi(\mathbf{k}_1, \ldots, \mathbf{k}_l) = \mathbf{m}$$

---

[17]Ibid.

for each set of natural numbers $\mathbf{k}_1, \ldots, \mathbf{k}_l$ was so deducible. It was this definition of general recursive functions that Kleene adopted from the lectures he attended at Princeton.

5.4. Kleene's Normal Form for General Recursive Functions

Kleene had acquired his definition of recursive functions from Gödel's lectures and this seems to be the only direct influence that Gödel had on Kleene at this time. However Gödel spoke to Church at length, and considerable discussion centered on the possibility that the general recursive functions encompassed all effectively calculable functions. Since Church had already formulated his thesis that the λ-definable functions were equivalent to the effectively calculable functions, it would be a vital piece of evidence in support of either hypothesis if it could be proved that the λ-definable functions were equivalent to the general recursive functions. Church also discussed these things with Kleene because in the academic year 1934-35, after he had completed his other papers, Kleene set out to prove this equivalence.

While working on this paper it became obvious that the easiest way to get the λ-definability of all general recursive functions was to produce the normal form for general recursive functions, so the work needed for the production of this normal form, and other associated results, was extracted and put into a separate paper, "General Recursive Functions of Natural Numbers" [86]. This paper was a spin off from his originally intended paper, "λ-Definability and Recursiveness" [87], and indeed it turned out to be the more interesting paper.

Kleene thus produced the first results in the theory of general recursive functions in his [86] paper. This paper is divided into two sections the first of which culminates in the proof of Kleene's normal form for general recursive functions, namely that every general recursive function is expressible in the form

$$\psi(\epsilon\, y\, [R(x_1, \ldots, x_n, y)]),$$

where $\psi(y)$ is a primitive recursive function and $R(x_1, \ldots, x_n, y)$ is a primitive recursive relation, with the restriction that

$$(\forall x_1, \ldots, x_n)\, (\exists\, y)\, R(x_1, \ldots, x_n, y)$$

and $\epsilon y\, [R(y)]$ means the least y for which $R(y)$ is true.

Kleene accomplishes this in stages. First he produces a purely formal definition for the class of general recursive functions. He abstracts all meaning from the terms, equations and definitions and treats the defining equations as a sequence of symbols. He then lists some formal operations on these symbols, which represent the rules of derivation given in Gödel's definition. Finally he arithmetizes this formalism to produce numbers that represent the formulae of the system, and in this way isolates the primitive recursive element of the definition. This process will now be described in more detail.

After his enumeration of all the formal symbols of his formalism Kleene lists the three operations that represent rules of procedure for passing from expression E to expression F in the formalism:[18]

$\mathbf{R}_1$: to replace A by $S^{x_1, \ldots, x_n}_{k_1, \ldots, k_n} A|$, where $x_1, \ldots, x_n$ are the numerical variables which occur in A, and $k_1, \ldots, k_n$ are numerals.[19]

$\mathbf{R}_2$: to pass from A and $\sigma(k_1, \ldots, k_s) = k$ to the result of substituting k for a particular occurrence of $\sigma(k_1, \ldots, k_s)$ in A, where $k_1, \ldots, k_s$, k are numerals.

$\mathbf{R}_3$: to pass from A and $B = C$ to the result of substituting C for a particular occurrence of B in A.

---

[18]See [**35**, p. 240].

[19]$S^{a_1, \ldots, a_n}_{b_1, \ldots, b_n} A|$ is defined to stand for the result of substituting b_i for a_i throughout A.

He writes $E \vdash_{1,2,3} F$ to denote passing from expression E to expression F via operations R_1, R_2 and R_3. These operations are the formal equivalents of Gödel's rules for evaluating a function via derived equations.

Operations R_1 and R_2 are indeed very similar to Gödel's derived equations, but since Kleene is now dealing with a purely formalized theory he specifies in Operation R_1 that he will replace symbols for variables with symbols for numerals whereas Gödel had spoken of replacing variables with natural numbers. Similarly Operation R_2 is the formal equivalent of the replacement in an equation of the calculated value of a function for that function itself.

Operation R_3 is an addition to Gödel's specified rules and it allows Kleene to substitute one equation with an equivalent one inside another equation. Operations R_1 and R_2 are used by Kleene to specify a definition of the class of general recursive functions similar to the Herbrand-Gödel definition, while operations R_1 and R_3 are used to give a second definition which, at first sight appears to be more wide ranging, but which Kleene proves is equivalent to the first definition.

Kleene had had plenty of experience in working with formally defined rules of procedure in his work on λ-definability so it is perhaps natural that he was the one to express the definition of general recursive functions in terms of a formalized theory.

5.5. The First Definition

The first of these definitions is the one that Kleene saw as equivalent to the Herbrand-Gödel definition and is stated this:[20]

---

[20]See [**35**, p. 240].

Definition 2a.[21] Given functional variables $\sigma_1, \ldots, \sigma_n$, let E_j^* denote the set of equations

$$\sigma_j(k_1, \ldots, k_{s_j}) = k$$

where k is the "value" of $\sigma_j(k_1, \ldots, k_{s_j})$ as presently defined. The functions $\sigma_1, \ldots, \sigma_n$ are *defined recursively* by the system of equations $(E_1, \ldots, E_n)$ if, for each $i(i = 1, \ldots n)$ E_i is a system of equations in $\sigma_1, \ldots, \sigma_i$, each of the form $\sigma_i(a_1, \ldots, a_{s_i}) = b$ where σ_i does not occur in $a_1, \ldots, a_{s_i}$, such that for each set of numerals $k_1, \ldots, k_{s_i}$ there is exactly one numeral k (called the value of $\sigma_j(k_1, \ldots, k_{s_j})$) for which

$$E_1^*, \ldots, E_{i-1}^*, E_i^* \vdash_{1,2} \sigma_i(k_1, \ldots, k_{s_i}) = k.$$

A function σ_n is *recursive* if there is an $(E_1, \ldots, E_n)$ of this description.

Perhaps a few words of explanation and an example will help in understanding this definition.

Firstly, E_i^* formally represents a set of equations giving the "value" of σ_i for all values of arguments $k_1, \ldots, k_{s_i}$. So that if σ_2 is defined by

$$\sigma_2(0) = 0$$
$$\sigma_2(\sigma_1(x_1)) = x_1$$

where σ_1 is the successor function S, then E_2^* will consist of equations like:

$$\sigma_2(S(0)) = 0$$
$$\sigma_2(S(S(0))) = S(0)$$

and so forth. That is, E_2^* contains σ_2 evaluated for all possible substitutions of numerals for numerical variables.[22]

---

[21]The preceding Definition 1 is the definition of the class of primitive recursive functions.
[22]The numerals are one of the expressions 0, $S(0)$, $S(S(0))$ ...

Secondly, each E_ι, is a system of equations containing only $\sigma_1, \ldots, \sigma_\iota$ where the left-hand side is of the form

$$\sigma_\iota(a_1, \ldots, a_{s_\iota})$$

and each a_j does not contain σ_ι. This restriction, that the left-hand side only contains one occurrence of σ_ι, is the same as that given by Gödel in [**54**].

Thirdly, the sequence of equations $(E_1, \ldots, E_n)$ contains the functions $\sigma_1, \ldots, \sigma_n$ an ordered in such a way that only a knowledge of the "values" of previous functions and the equations E_ι themselves are needed to calculate any function σ_ι that is

$$E_1^*, \ldots, E_{\iota-1}^*, E_\iota$$

is sufficient to calculate $\sigma_\iota(k_1, \ldots, k_{s_\iota})$ uniquely for all numerals $k_1, \ldots, k_{s_\iota}$ This again conforms with Gödel's intention for his definition.

The whole definition still incorporates Herbrand's original idea of sequences of equations $(E_1, \ldots, E_n)$ containing functions substituted into one another in a general fashion. But the restrictions on this substitution, namely:

    i) the form of the operations R_1 and R_2
    ii) the fact that only $E_1^*, \ldots, E_{\iota-1}^*, E_\iota$ are required to evaluate σ_ι.

are Gödel's restrictions.

Finally, this definition can be thought of as recursively defining n functions simultaneously. For example:

$$E_1 \vdash_{1,2} \sigma_1(k_1, \ldots, k_{s_1}) = k \text{ giving all } E_1^*$$
$$E_1^*, E_2 \vdash_{1,2} \sigma_2(k_1, \ldots, k_{s_2}) = k \text{ giving all } E_2^*$$

and so forth. In practice, though, it is the last function that is explicitly being defined. This function is now represented by the principal function letter σ_n

in the definition, and is said to be defined recursively in terms of the other auxiliary function letters.

5.6. An Example of the First Definition

With the exception of the constant and identity functions, all the functions used in the following example are denoted by symbols with different indices, and in particular σ_1 stands for the initial function S, the successor function. The following sequence of equations $(E_1, \ldots, E_5)$ defines the function σ_5. An explanation of each of the functions is given alongside the definition:

E_1	$\sigma_1(x_1) = S(x_1)$	successor
E_2	$\sigma_2(0) = 0$	predecessor
	$\sigma_2(\sigma_1(x_1)) = x_1$	
E_3	$\sigma_3(x_1, 0) = x_1$	modified difference
	$\sigma_3(x_1, \sigma_1(x_2)) = \sigma_2(\sigma_3(x_1, x_2))$	
E_4	$\sigma_4(x_1, 0) = x_1$	addition
	$\sigma_4(x_1, \sigma_1(x_2)) = \sigma_1(\sigma_4(x_1, x_2))$	
E_5	$\sigma_5(x_1, x_2) = \sigma_4(\sigma_3(x_1, x_2), \sigma_3(x_2, x_1))$	modulus

Hence $E_1, \ldots, E_5$ is said to define σ_5, that is

$$E^*{}_1, \ldots, E^*{}_4, E_5 \vdash_{1,2} \sigma_5(k_1, k_2) = k$$

for all k_1 and k_2.

5.7. The Second Definition

The second definition of the class of general recursive functions given by Kleene is:[23]

---

[23]See [**35**, p. 241].

Definition 2b. The functions $\sigma_1, \ldots, \sigma_n$ an are *defined recursively* by E, if E is a system of equations in $\sigma_1, \ldots, \sigma_n$ such that for each $\imath, \imath = 1, \ldots, n$ and each set of numerals $k_1, \ldots, k_{s_\imath}$ there is exactly one numeral k, (called the value of $\sigma_\imath(k_1, \ldots, k_{s_\imath})$) for which

$$E \vdash_{1,3} \sigma_\imath(k_1, \ldots, k_{s_\imath}) = k.$$

A function σ_n is *recursive* if there is an E of this description.

This definition uses the less specific operation R_3 and hence appears to give a wider class of functions, but Kleene later proves it to be equivalent to definition $2a$. At this stage he proves that it is not less general than definition $2a$ and then proceeds to use it for his next task, Gödel numbering.

Since operation R_3 allows the substitution of one expression, B, for another, C, in the expression, A, without B necessarily being $\sigma(k_1, \ldots, k_s)$ and C necessarily being k, then it is possible to break operation R_3 down into simpler steps for Gödel numbering more easily than operation R_2. This can be accomplished by considering in turn the first, second, third, $\ldots$, symbol of A as the start of expression B. Hence Kleene uses definition $2b$ rather than definition $2a$ in this arithmetization stage.

Kleene utilizes the technique of Gödel numbering to represent all the expressions, terms, formulæ and equations of his formal system with numerical relationships, all of which he demonstrates to be primitive recursive. Having broken operations R_1 and R_3 into very small steps he can easily code these operations into numerical relationships by gradually combining simpler ones. In this way he proves that the use of operations R_1 and R_3 is equivalent to the use of primitive recursive operations on the numbers that code the expressions of his formalism, that is, the rules can be represented by primitive recursive functions. Using these primitive recursive functions and relations he then proves a theorem to the effect:

If z is the number for a system of equations Z, then there is
a primitive recursive function $H(z, m)$ which enumerates with
repetitions all the numbers corresponding to equations that
can be derived from equations Z by use of the operations R_1
and R_3.

Hence there is a primitive recursive enumeration of the generated equations.

Now suppose E is a system of equations defining the function $\phi(x_1, \ldots, x_n)$
recursively under definition 2b. Then for all $(x_1, \ldots, x_n)$ there exists a value
x' for $\phi(x_1, \ldots, x_n)$. That is, $\phi(x_1, \ldots, x_n) = x'$ must be one of the equations
generated from E by operations R_1 and R_3. Now if E has Gödel number e
then $H(e, m)$ will enumerate with repetitions all the numbers of the equations
generated from E using operations R_1 and R_3. So there must be at least one
value of m, y say, such that

$H(e, y)$ is the Gödel number of $\phi(x_1, \ldots, x_n) = x'$.

By Kleene's previous results correlating his formalism with primitive recursive
functions and relations it is possible to prove that:

i) There exists a primitive recursive relation $R(x_1, \ldots, x_n, y)$ which is
true precisely if $H(e, y)$ is the Gödel number of $\phi(x_1, \ldots, x_n) = x'$ So,
since there must be at least one y, then

$$(\forall x_1, \ldots, x_n)\, (\exists y)\, R(x_1, \ldots, x_n, y)$$

must be true.

ii) There is a primitive recursive function $\psi(y)$ such that $\psi(y) = x'$. That
is, the value of $\phi(x_1, \ldots, x_n)$ can be found by applying the function ψ
to one of the values of y such that $R(x_1, \ldots, x_n, y)$ is true.

By taking the least value of y this gives x' equal to:

$$\phi(x_1, \ldots, x_n) = \psi(\epsilon\, y\, [R(x_1, \ldots, x_n, y)])$$

which is Kleene's normal form, Theorem IV.

Hence any function recursive in the sense of definition 2b (or definition 2a, since, if anything, definition 2a appears less general) is expressible in the form:

$$\psi(\epsilon\, y\, [R(x_1, \ldots, x_n, y)])$$

where $\psi(y)$ is a primitive recursive function, $R(x_1, \ldots, x_n, y)$ is a primitive recursive relation and the function $\epsilon\, y$ is the least number operator.

This significant result pinpoints the essential difference between general recursive functions and primitive recursive functions; namely, the unbounded use of the function $\epsilon\, y$. Since

$$\epsilon\, y\, [y \leq \chi(x_1, \ldots, x_n)\, \&\, R(x_1, \ldots, x_n, y)]$$

is primitive recursive by [51], it must be the lack of a bound on the value of y that makes a function general recursive rather than primitive recursive. Thus the operation of searching indefinitely through the natural numbers for one satisfying a primitive recursive relation is all that needs to be added to substitution and primitive recursion to obtain a general recursive function.

It should be noted that the function $\epsilon\, y$ used here is defined to be zero if no y exists. Later, in "On Notation for Ordinal Numbers" [89], Kleene uses a definition of the least number operator that is undefined if the sought value of y does not exist and hence defines partial recursive functions.

This, in fact, is the way the $\mathfrak{p}$ function, the equivalent function used in his work on λ-definability, is defined. With partial functions and the alternate definition for the least number operator it is no longer necessary to assume that

$$(\forall x_1, \ldots, x_n)\, (\exists y)\, R(x_1, \ldots, x_n, y)$$

is true. On analyzing the differences between the primitive recursive function

$$\epsilon \, y \, [y \leq \chi(x_1, \ldots, x_n) \, \& \, R(x_1, \ldots, x_n, y)]$$

and the general recursive function

$$\psi(\epsilon \, y \, [R(x_1, \ldots, x_n, y)])$$

it can be seen that the unboundedness of y in the second definition implies that the length of computation of a general recursive function must grow faster with the values of the arguments than the value of any primitive recursive function or else that primitive recursive function could be used as a bounding function $\chi(x_1, \ldots, x_n)$ which would reduce the general recursive function in the second definition to a primitive recursive function.

This result, incorporating the one unbounded operator, seems an incredible simplification compared with the original format of a general recursive function. Yet it can be seen as a natural development from Kleene's previous work, since, as Kleene states in a letter to the author of 28 July 1977, the least number operator was the natural descriptive operator for working with the positive integers or the natural numbers. Also Kleene had already known for nearly two years that the class of λ-definable functions, which he was then in the process of proving equivalent to the general recursive functions, was closed under primitive recursion, composition and this least number operator.

Kleene concludes the first section of his paper by proving the second half of his equivalence between definition $2a$ and definition $2b$, and then proving two theorems on recursive enumerability.

A class is defined to be recursively enumerable if there exists a recursive function f such that $f(n)$ generates the whole class. Also, using Gödel's method of defining primitive recursive classes and relations, Kleene defines a class to be recursive if its representing function is recursive.[24] Kleene now proves that if a class is infinite and recursive then the function that enumerates the members without repetition and in order of magnitude is recursive, that is, an infinite class is recursively enumerable without repetitions and in order of magnitude if and only if it is recursive.

Finally he proves that the recursive enumerability with repetitions of an infinite class implies its recursive enumerability without repetitions. Hence if a non-recursive class is recursively enumerable, the enumeration cannot be in order of magnitude.

The concept of recursive enumerability originated with Kleene.[25] As he states in a letter to the author of 28 July 1977, general recursive functions had not been known of long enough for anyone else to consider the idea and although the equivalent question of which classes of numbers were given by the range of a λ-definable function could have been considered before then, it had not been.

5.8. Other Developments

In the second half of his paper Kleene considers various other questions in the theory of recursive functions. He inquires whether it is possible to determine which systems of equations will define a function recursively or, equivalently, whether the Gödel numbers of these systems are recursively enumerable. He examines the possibility of there being nonrecursive classes and functions and also he produces a general undecidability result for a large class of formal logics.

---

[24]The term recursive is used to signify general recursive from now on. Since primitive recursive functions are just special cases of general recursive functions this implies a special case result if the function or class happens to be primitive recursive.

[25]In 1944 Post [**129**] tidied up this definition by including the empty set as also recursively enumerable.

His first task is to investigate the problem of finding systems of equations which will define a function recursively. He comes to the conclusion that there can be no constructive procedure for determining these systems, where, by a constructive procedure, he means a recursive one.

To solve this problem he utilizes the correspondence between the systems of equations E and their Gödel numbers e and he asks, instead, the equivalent number-theoretic question of which numbers will define functions recursively. He proves that these numbers are not recursively enumerable. This result is not unexpected since, if it is possible to enumerate recursively these Gödel numbers, then the diagonal process can be applied to generate yet more general recursive functions.

As a first step in this task Kleene proposes a third definition for general recursive functions, definition 2c. He defines a primitive recursive relation $T_n(z, x_1, \ldots, x_n, y)$ which is true if the yth number enumerated by the function $H(z, m)$ corresponds to the equation

$$\phi(x_1, \ldots, x_n) = x'$$

for some function ϕ of the arguments $x_1, \ldots, x_n$. That is, if the yth equation generated from the system of equations Z with Gödel number z is of the form

$$\phi(x_1, \ldots, x_n) = x'$$

for some ϕ, then $T_n(z, x_1, \ldots, x_n, y)$ is true. He can then state definition 2c which says that a function $\phi(x_1, \ldots, x_n)$ is recursive if there exists a number e such that

$$(\forall x_1, \ldots, x_n) \, (\exists y) \, T_n(e, x_1, \ldots, x_n, y)$$

is true.

This definition is completely in terms of Gödel numbers since it requires the existence of a number e such that for each $x_1, \ldots, x_n$ $H(e, m)$ will eventually generate a number of the form $\phi(x_1, \ldots, x_n) = x'$; only indirectly does the definition refer to a system of equations E.

The definition continues by specifying that the value of the function $\phi(x_1, \ldots, x_n)$ will be the one given by the first equation generated by $H(e, m)$. Now the function $H(e, m)$ enumerates with repetitions and consequently the value x' can occur many times for many values of m, which is why the least value function is again needed. But, not only that, this definition does not exclude the possibility that $H(e, m)$ may generate an equation of the form

$$\phi(x_1, \ldots, x_n) = x''$$

for some later value of m. Hence the system of equations E, for which e is the Gödel number, could be a system that determines a multiple valued function.

Definition $2c$ only requires that $\phi(x_1, \ldots, x_n)$ should have the first value in the list given by $H(e, 0), H(e, 1), H(e, 2), \ldots$. So the system E used here could be inadmissible as a system of equations under definition $2a$ or $2b$, since these definitions require that a single valued function shall be defined.

Despite this, Kleene is able to prove that the class of functions defined by definition $2c$ is equivalent to the class of functions defined by definition $2b$ or $2a$.

To show that the class of functions given by definition $2b$ or $2a$ is included in the class given by definition $2c$ he notes that for any system of equations E used in definition $2b$ or $2a$, the Gödel number e of this system can be used in definition $2c$.

To show that the class of functions defined under definition $2c$ is not greater than the class of functions defined by definitions $2a$ and $2b$, despite the different types of systems of equations E allowed in the definition, he remarks that any function recursive under definition 2c is expressible in the form

$$\psi(\epsilon\, y\ [T_n(e, x_1, \ldots, x_n, y)])$$

where $\psi(y)$ is a primitive recursive function and $T_n(e, x_1, \ldots, x_n, y)$ is a primitive recursive relation, and is therefore of the correct form to be recursive under definition 2b and hence 2a.

Armed with this new number-theoretic definition for recursive functions, Kleene can answer the question as to whether the numbers that define recursive functions are recursively enumerable. He does this in the next two theorems where, for convenience, he only considers functions of one variable. That is, he only uses $T_1(z, x, y)$ rather than $T_n(z, x_1, \ldots, x_n, y)$

He considers an arbitrary recursive function $\theta(x)$, each of whose values are Gödel numbers of a system of equations E_x that defines a recursive function ϕ_x of one variable. Therefore the functions ϕ_x represent a sequence of recursive functions whose defining Gödel numbers form a set which is recursively enumerated by the function $\theta(x)$. He then applies the diagonalizing procedure to this sequence to define a new recursive function $\eta(x)$ by:

$$\eta(x) = \phi_x(x) + 1.$$

He proves that for this function:

$$(\forall x)\ (\exists y)\ T_1(f, x, y)\ \text{and}\ (\forall x)\ \theta(x) \neq f.$$

That is if the defining Gödel number for the function $\eta(x)$ is the number f then f does not appear amongst the values of the enumerating recursive function $\theta(x)$.

It is well known that the diagonal argument applied to any listing of primitive recursive functions leads to the definition of a non-primitive recursive function, but Kleene has now proved that the diagonal argument applied to any list of recursive functions whose defining Gödel numbers are recursively enumerable leads to the definition of a function not contained in that list but that is still general recursive. Hence the totality of numbers which define functions recursively cannot be enumerated by any recursive function $\theta(x)$ by Theorem XI.

One consequence of this result is that the class of numbers z such that

$$(\forall x)\,(\exists y)\,T_1(z, x, y)$$

is not recursive. That is, the class of numbers which define functions recursively is not recursive or else it could be recursively enumerated in contradiction to Theorem XI. This gives the first example of a nonrecursive class.

The existence of this non-recursive class, or the non-recursive enumerability of the numbers that define recursive functions, is now used by Kleene to prove the existence of undecidable number-theoretic propositions in certain formal logics.

The first result of this type had been, of course, produced by Gödel in 1931, but here Kleene proves the existence of the undecidable proposition by utilizing reasoning of the following form. If there are no undecidable propositions in the logic then that logic can be used to construct a recursive enumeration of all the numbers that will define functions recursively and this will contradict Theorem XI.

The formal logics, S under consideration in this theorem have to obey certain restrictions:

i) The proposition $(\forall x)\,(\exists y)\,T_1(z, x, y)$ can be expressed in the symbolism of S by formulæ A_z, that is

$$(\forall x)\,(\exists y)\,T_1(N, x, y)$$

is expressed by some A_N.

ii) The forumuæ of S can be assigned numbers so that:

   a. distinct formulæ A_z get distinct numbers a_z

    b. the class of numbers of axioms is recursive

    c. the relationship between numbers that represent forumlæ involved in the rule of immediate consequence is recursive

    d. the class of numbers a_z of forumuæ A_z is recursive

    e. the function that yields the number z from the number a_z of the formula A_z is recursive

  iii) If the formula A_z is provable in the system S, then the equivalent proposition $(\forall x)\,(\exists y)\,T_1(z, x, y)$ outside the logic is to be true.

These restrictions allow Kleene to construct a function $\theta(w)$ that will enumerate all the numbers z for which A_z is provable. Invoking the last restriction, this is equivalent to

$$(\forall x)\,(\exists y)\,T_1(\theta(w), x, y)$$

is true for all w.

But, by Theorem XI, for any function $\theta(w)$ there must exist a number f such that

$$(\forall x)\,(\exists y)\,T_1(f, x, y)$$

is true and yet $\theta(w) \neq f$.

Hence the proposition $(\forall x)\,(\exists y)\,T_1(f, x, y)$ is true yet its equivalent formula A_f is not in the enumeration of provable formulae and therefore is not provable in the logic S.

So there exists a formula in the logic which represents a true proposition but which is also unprovable.

This theorem does not actually give an undecidable proposition in the sense of showing that neither A_f nor $\neg A_f$ is provable in the logic, but it does show that if the logic is correct[26] then it is not complete.[27]

Kleene indicates in a footnote how this can be converted into a proper undecidability result and he amplifies this in Kleene 1943 where the theorem is designated the generalized Gödel theorem.

Because Church's thesis equates recursive with effective and since Kleene had already accepted Church's thesis at the time of writing this paper, it can be seen that most of his restrictions on the logic S are specifying effective processes. He requires an effective method of calculating certain functions and an effective means of deciding whether a formula represents an axiom ar not and so on. The restrictions are all, therefore, desirable properties of any logic, and the last restriction, requiring the logic to be correct, is the main purpose of constructing the logic in the first place. This theorem shows that the other half of this equivalence between provable formulae and true statements is not provable. In other words, that all true statements should be provable within the formal system is not provable.

The final part of Kleene's paper consists of several theorems relating to non-recursive functions and non-recursive classes.

i) The function $\epsilon\, y\, [T_1(x,x,y)]$ is non-recursive. This is the first non-recursive function given in Kleene's paper, though others are given by Church in [23]. It is non-recursive despite being constructed out

---

[26] A system of logic is said to be correct if, under the intended interpretation, every provable formula is true – that is restriction (iii) in the above list of conditions on the logic S.

[27] A system of logic is said to be complete (in the meaning intended here) if every true statement is represented in the logic by a formula which is provable.

of a primitive recursive relation $T_1(x, x, y)$ with the use of only one quantifier.

ii) The class $(\exists y)\ T_1(x, x, y)$ is non-recursive, yet it is recursively enumerable. Hence, by previous results, the enumerability must not be in order or else the class would be recursive.

iii) The class $\neg(\exists y)\ T_1(x, x, y)$ is not recursively enumerable. Kleene proves this result non-constructively by showing that, if this class if recursively enumerable then, together with the fact that $(\exists y)\ T_1(x, x, y)$ is recursively enumerable, we could deduce that the $(\exists y)\ T_1(x, x, y)$ is recursive in contradiction to (ii) above.

The method used in the proof of Kleene's result appears later as a full theorem in Post's paper [129] in the form:

A class is recursive if and only if it and its complement are recursively enumerable.

Kleene claims, in a footnote to the 1936 paper, that he has actually got a constructive proof of the fact that, for certain recursive relations $R(x, y)$, the class $\neg(\exists y)\ R(x, y)$ is not recursively enumerable. A constructive proof that $\neg(\exists y)\ T_1(x, x, y)$ was not recursively enumerable was circulated privately by Kleene in [87] and it can be found in [35, p. 253], where it has been included as an Addendum to the reprint of Kleene's paper contained therein.

iv) Finally, Kleene proves Theorem XVII which shows that, for any recursive relation

$$R(x_1, \ldots, x_n, y_1, \ldots, y_m),$$

there is a number e such that:

$$(\forall x_1, \ldots, x_n)\ (\exists y_1, \ldots, y_m)\ R(x_1, \ldots, x_n, y_1, \ldots, y_m) \equiv (\forall x)\ (\exists y)\ T_1(e, x, y)$$

where T_1 is, of course, primitive recursive.

So whether or not a primitive recursive relation holds is proved equivalent to whether or not a function defined by some number e is recursive.

Since the question of which numbers define functions recursively is not effectively solvable, then this theorem shows that the problem of whether or not a recursive relation holds is not effectively solvable either. Kleene points out that this theorem is similar to Theorem 18IV in [85], which states that the solvability of a given formula is equivalent to whether or not a certain function is λ-definable. Kleene had, at this stage, already proved that the classes of recursive functions and λ-definable functions were the same, so this similarity is not surprising.

Also this theorem gives, for values of $e = 0, 1, 2, 3 \ldots$ an enumeration with repetitions of all predicates of the form

$$(\forall x_1, \ldots, x_n) (\exists y_1, \ldots, y_m) \, R(x_1, \ldots, x_n, y_1, \ldots, y_m)$$

where R is general recursive. In his 1943 paper Kleene uses this theorem, called the enumeration theorem, at an earlier stage to derive most of the last theorems of the 1936 paper more directly.

5.9. Conclusion

In this chapter we have examined the two main strands of development of recursive functions during the early 1930's, a period of great advance in recursive function theory. Firstly, in a series of papers on recursive functions, Rózsa Péter analyzed and extensively developed the theory of primitive recursive functions, the definitive account of this work being found in her 1951 book. Secondly, the extremely important class of general recursive functions was introduced by Herbrand and Gödel and developed by Kleene.

Herbrand initiated the generalization of the recursive definition by ignoring the iterative part of the usual recursive definition and concentrating on the formal equation side of that definition; following him Gödel considerably modified this definition to secure effectiveness. As we have seen, although Gödel introduced the functions with the idea that they might represent all the effectively calculable functions, he was not sure that they represented all possible recursions and so remained unconvinced of Church's thesis until after Turing's introduction of his computable functions to be taken up in Chapter 7.

Kleene took Gödel's definition of general recursive functions and adapted it on the formal side. His first two definitions, like Gödel's definition, consisted of a specification of the form of the equations allowed in the definition, and a description of the nature of the operations allowed in the computation of the values of the function. In each of these definitions there was included the restriction that, for each set of arguments, the computation should yield a unique number as value. This was achieved in the third definition, definition 2c, in Kleene's paper by taking the first value generated by the function of the Gödel numbers involved in the definition. Hence in each definition there was always the assumption that a value would be produced eventually, since the definitions were prefaced by some statement such as:

$$(\forall x_1, \ldots, x_n)(\exists y)R(x_1, \ldots, x_n, y).$$

This slightly artificial definition, giving a class of total functions, was later changed in [89] into a definition of a class of partial functions. In this new definition the requirement that the computation process should terminate with a unique equation of the form:

$$\phi(k_1, \ldots, k_n) = k$$

was omitted and the function involved was a partial function being defined over a subset, possibly null or total, of the n-tuples of natural numbers.

In his paper, Kleene had indicated the precise nature of the extension of the general recursive functions over the primitive recursive functions by giving his normal form theorem. This showed that the bound on the value of y in the least y function was the only essential difference between these classes of functions. This development was a significant one and in [**94**] he explains how he thought of this normal form theorem involving primitive recursions with explicit definitions and the least-number operator:[28]

> I had been preconditioned by my work on λ-definability to think in terms of these elements. Thus I had confirmed in my PhD thesis that all primitive recursions (as well as explicit definitions) can be effected in the λ-calculus, and likewise the least-number operator. Part of the project for my 1936 paper was to prove that every general recursive function is λ-definable; so I could not help but reflect that I could do that if I could get every general recursive function by a combination of primitive recursions (with explicit definitions) and least-number operations.

In the second part of [**86**] Kleene devoted himself to the problem of which systems of equations defined functions recursively, or the equivalent problem, using his third definition, of which numbers defined functions recursively. He came to the conclusion that there was no effective means of determining these systems of equations or these numbers. He argued that this was to be expected since, if effective meant recursive as Church's thesis specified, then the existence of any recursive method of determining the appropriate systems of equations or numbers would allow the generation, by a diagonal process, of other more generally recursive functions.

Kleene also used Church's thesis, plus his normal form theorem and the subsequent definition $2c$ of general recursive functions in terms of Gödel numbers, to produce the first generalized version of Gödel's undecidability theorem.

---

[28]See [**94**, p. 60].

Kleene's theorem, being expressed in terms of recursive functions and recursively enumerable classes, made it clear to which systems results like Gödel's would apply assuming that Church's thesis, equating recursive with effective, was accepted. Provability corresponded to recursive enumerability and the recursive nature of the formal system indicated the effective nature of the construction of the formal system.

As previously noted, all of Kleene's paper was written after Church's thesis had been formulated and after Kleene's acceptance of it, even though Church's paper was not published until later. As a result most of Kleene's theorems should be read with Church's thesis in mind. Both the λ-definable functions considered in Chapter 4 and the general recursive functions considered in this chapter gain in significance and power due to Church's thesis. One of the major reasons for accepting Church's thesis is the fact that these two sets of functions are identical, and it is this proof that we shall consider in the next chapter, along with other evidence for Church's thesis.

CHAPTER 6

Church's Thesis

6.1. Introduction

The principal objective of this chapter is to discuss Church's thesis and some of its early consequences. During the years before its enunciation, in 1934, the need for some formal equivalent to the imprecise and intuitive concept of effectively calculable functions had become more apparent and, indeed, more urgent since many problem were arising in mathematics whose solutions required the definition of an effectively calculable function. Church, recognizing this need, began his 1936 paper with a brief discussion of some examples of effectively calculable functions from the domains of elementary number theory and topology.

Hilbert had already recognized the need for a more precise notion of effectively calculable functions in his work of the 1920's. He realized that the primitive recursive functions were certainly contained in that class but that other functions, for example Ackermann's function, were also effectively calculable. Many particular functions could be proved effectively calculable but with the consideration of more general problems, such as decision problems for logics, the need arose for a method of characterizing all effectively calculable functions. Results obtained by Church, Kleene and Rosser in the early 1930's led Church, around February or March 1934, to posit that the class of effectively calculable functions was coexistent with the class of λ-definable functions.[1]

While Gödel was visiting Princeton he introduced into his lectures, around April 1934, the concept of the class of general recursive functions and Church recognized that this class also seemed to be equivalent to the class of effectively calculable functions. Both Church and Kleene set to work on the proof of the equivalence between the general recursive functions and the λ-definable functions. The final statement of Church's thesis, which was first given in [**23**], was that both the classes of λ-definable functions and general recursive functions should be equated with the class of effectively calculable functions.

The second section of this chapter will deal with this important equivalence proof between the λ-definable functions and the general recursive functions. If this proof had not been successful, then Church's thesis could not have been stated in the form it was. In fact in the first published announcement of Church's thesis, in the abstract of Church's 1936 paper,[2] Church only mentioned general recursive functions and not λ-definable functions. Kleene has expressed the opinion that, since Church originally conceived his thesis in terms of λ-definable functions, it seems most likely that the omission of any reference to them now was because Church had only definitely proved that λ-definability implied recursiveness and not the converse. Indeed, in his 1936 paper, Church only claimed to have proved this half. Consequently it could be that Church stated his thesis using what he thought, at the time, was the most general class of functions.[3]

---

[1]This was the date when Church asserted the thesis definitely to Kleene.

[2]This abstract is contained in the Bulletin of the American Mathematical Society of May 1935 (Volume 41 Pp 332-333) and was received on 22 March 1935.

[3]In fact there seems to be a certain amount of confusion on Church's part, early in 1935, as to whether or not the whole of the equivalence proof had been produced. He mentioned in [**23**, footnote 16] a proof by Rosser and be may have thought, originally, that this represented a full proof of the "recursive implies λ-definable" part of the equivalence. It was certainly true that everyone at Princeton was convinced that the proof would eventually be found.

To prove the equivalence between the λ-definable functions and the general recursive functions various new recursive results were needed, as well as some new definitions and general modifications of the old definitions. At the same time the important concept of normal form for λ-definable functions was emphasized and the concept of recursion was extended to include functions and operations on well-formed formulae, as well as on natural numbers. Church also gave the definition of potential recursion.

All these results were contained in the two papers, Church's "An Unsolvable Problem of Elementary Number Theory" [23] and Kleene's "λ-Definability and Recursiveness" [87]. Kleene's paper concentrated on the detailed proof of the equivalence between λ-definable functions and general recursive functions and extended this result to some more general functions. Church's paper concentrated on the statement and evidence for his thesis and the developments leading from it. He just listed the major results from the equivalence proof and showed how to use them to build up evidence for his thesis and subsequent theorems. The research for both these papers was carried out during the same period. Kleene, as discussed in chapter 5, was also extracting all the information pertaining to his normal form for general recursive functions and putting it into another paper, "General Recursive Functions of Natural Numbers" [86]. It was Kleene's possession of his normal form theorem for general recursive functions that allowed him to complete the "recursive implies λ-definable" part of the equivalence proof.

All three papers were completed in the first half of 1935. Church was the first to send off an abstract of his work but Kleene was actually the first to finish and send off all the work. In fact Kleene completed and sent off both his papers at the end of June, giving copies to Church. Church, on the other hand, must have sent off his final copies slightly later since he informed Kleene, in a letter of 6 July, that he had made a number of changes to his paper since they had last met in late June.[4]

In his papers Kleene made considerable use of the results from his PhD thesis and developed an impressive expertise with the Gödel numbering technique. Together Kleene's two papers contain a wealth of information on general recursive functions. However, the major importance of both these functions and the λ-definable functions results from the proposal to equate them to the class of effectively calculable functions that is contained in Church's outstanding paper.

6.2. λ-Definable vs. General Recursive

Both Church, in "An Unsolvable Problem of Elementary Number Theory", and Kleene, in "λ-Definability and Recursiveness", investigate the proof of the equivalence between the class of λ-definable functions and the class of general recursive functions. Church only gives an indication of the proof by stating the major results required for the equivalence theorem, and then, in some cases, discussing briefly how the proof of that result could be achieved. Both of his classes of functions are defined for the domain of positive integers, in keeping with his previous work on λ-definable functions. Kleene, on the other hand. gives full details of the equivalence proof since it was he who actually produced the whole proof of equivalence and considers his functions defined on the class of natural numbers. This use of natural numbers conforms with Kleene's other paper on recursive functions [86]. As he points out on page 343 of the "λ-Definability and Recursiveness" paper [87], the choice of natural numbers or

---

[4]The exact details and dates for Church's and Kleene's papers are as follows. Church's abstract was received for publication on 22 March and was published in May. Church spoke to the American Mathematical Society about his paper on 19 April. Church probably sent his completed paper for publication in early July. Abstracts of Kleene's two papers were received by the American Mathematical Society on 1 July. The actual papers were received by the *Duke Mathematical Journal* [87] on 1 July, and *Mathematische Annalen* [86] on 7 July.

positive integers is non-essential to the theory since either proof can easily be converted into an equivalent proof using the other set of numbers.

Both Church's and Kleene's papers start with a description of the structure of λ-notation and the rules for the operation of conversion between well-formed formulæ. These definitions are very clear and succinct and summarize, in a much better form, much of the previous work on these functions. Church also gives a clear account of the definition of a function by recursion. while Kleene just refers to his *Mathematische Annalen* paper. It is worth noting that the term "λ-definable functions" was, in fact, first used in this paper by Church.

Church gives the three axioms for conversion between well-formed formulæ much as they had appeared before and then proceeds to describe the type of conversion referred to as a reduction[5] and to define normal form:[6]

> A formula is said to be in normal form if it is well-formed and contains no part of the form $\{\lambda x[M]\}(N)$.

He also defines principal normal form but more will be said about this later. He then specifies what is meant by a function of a positive integer being λ-definable:[7]

> A function F of one positive integer is said to be λ-*definable* if it is possible to find a formula **F** such that, if $F(m) = r$ and **m** and **r** are the formulas for which the positive integers m and r stand according to our abbreviations introduced above, then $\{\mathbf{F}\}(\mathbf{m})$ conv **r**.

---

[5]A reduction contains exactly one application of Operation II and no applications of Operation III - as defined previously. See page 95 of this book.

[6]See [**23**, p. 348].

[7]See [**23**, p. 349].

He provides a similar definition for a function of several arguments.

Then, by quoting three theorems, two of which are from another paper, he indicates some of the important properties of normal form:[8]

> Theorem I: If a formula is in normal form, no reduction of it is possible.
>
> Theorem II: If a formula has a normal form, this normal form is unique to within applications of Operation I, and any sequence of reductions of the formula must (if continued) terminate in the normal form.
>
> Theorem III: If a formula has a normal form, every well-formed part of it has a normal form.

Note that Theorem II does not give any limit to the number of reductions needed, so, just because a sequence of reductions has not terminated after a certain number of applications of Operation II, it does not mean that the formula has not got a normal form. But if a formula does have a normal form then the sequence must terminate at some stage. Church finishes this section by stating:[9]

> It is clear that, in the case of any λ-definable function of positive integers, the process of reduction of formulas to normal form provides an algorithm for the effective calculation of particular values of the function.

---

[8]See [**23**, p. 348]. The first theorem is obvious, while the second and third come from [**28**].

[9]See [**23**, p. 349].

This statement again indicates the importance of normal form and this is further demonstrated when we consider the latter part of Church's paper and his consideration of effective methods for solving various decision problems.

Kleene, in his paper, gives a new but equivalent form for the definition of the operation of conversion between well-formed formulæ:[10]

> We introduce an equivalence relation **A** conv **B**, or **A** is *convertible* into **B**, between well-formed formulas, which is defined to be the relation of least domain which is
> (1) reflexive
> (2) symmetric
> (3) transitive
> and has further the properties that
> (4) if **A** conv **B**, then
>
> $$\{\mathbf{C}\}(\mathbf{A}) \text{ conv } \{\mathbf{C}\}(\mathbf{B})$$
> $$\{\mathbf{A}\}(\mathbf{C}) \text{ conv } \{\mathbf{B}\}(\mathbf{C})$$
> $$\lambda\mathbf{x}[\mathbf{A}] \text{ conv } \lambda\mathbf{x}[\mathbf{B}]$$
>
> (5) if the proper symbol **y** does not occur in **A**, then
>
> $$\lambda\mathbf{x}[\mathbf{A}] \text{ conv } S_{\mathbf{y}}^{\mathbf{x}}\lambda\mathbf{x}[\mathbf{A}]|$$
>
> (6) if **x** and the free symbols of **N** are not bound symbols of **M**, then
> $$\{\lambda\mathbf{x}[\mathbf{M}]\}(\mathbf{N}) \text{ conv } S_{\mathbf{N}}^{\mathbf{x}}\mathbf{M}|$$
>
> .

He explains that this equivalence relation **A** conv **B** corresponds to the relation of "equality in meaning" and then comments that:[11]

---

[10]See [**87**, p. 371].
[11]See [**87**, p. 341–342].

> "...the expression that we abbreviated to $\mathbf{F}(\mathbf{N}_1, \ldots, \mathbf{N}_n)$ represents the value of $\mathbf{F}$ (considered as a function of n variables) for the set of arguments $\mathbf{N}_1, \ldots, \mathbf{N}_n$; and the expression which we abbreviate $\lambda\mathbf{x}_1, \ldots, \lambda\mathbf{x}_n.\mathbf{M}$ represents the function which $\mathbf{M}$ is of $\mathbf{x}_1, \ldots, \mathbf{x}_n$."

This explanation furnishes us with a very clear understanding of the intended meaning of the λ symbolism and the operation of conversion. Indeed it is the first clear explanation given in print.

Church, in his 1936 paper, gives a lucid exposition of Kleene's formal version of the class of general recursive functions of Herbrand-Gödel, augmented by some excellent examples. The definition of general recursion is extended to functions defined on the class of well-formed formulae by both Kleene and Church, but we will consider this in more detail after we have outlined the details of the main equivalence result.

6.2.1. The Equivalence Theorem. The proof of the equivalence between the classes of λ-definable functions and general recursive functions is produced by Kleene in two stages.

First, Kleene proves that all the general recursive functions in the sense of Herbrand-Gödel are λ-definable. The major part of this proof had already been achieved by Kleene in his 1935 paper, where he had shown that the class of primitive recursive functions, and certain extensions of that class, were λ-definable. The proof in this 1936 paper proceeds on the same lines. Kleene first shows that the successor, zero and identity functions are λ-definable and hence that all the initial functions are λ-definable, since the successor and zero together would give any constant function. Also he proves that the definition of a function in terms of previous functions by substitution can be written in λ-notation. Then, using results on the predecessor function, maximum and minimum functions, and combinations, he succeeds in proving that any function,

L, defined by the primitive recursive schema:

$$L(0, x_2, \ldots, x_n) = G(x_2, \ldots, x_n)$$
$$L(S(y), x_2, \ldots, x_n) = H(y, L(x_2, \ldots, x_n), x_2, \ldots, x_n)$$

can be λ-defined in terms of two given formulae with no free symbols, **G** and **H** that λ-define the functions G and H. The condition concerning free symbols was included because the well-formed formulæ that represent the natural numbers, namely $\lambda fx.f(x), \lambda fx.f(f(x)), \ldots$, have no free symbols and it was known that interconvertible formulæ have the same free symbols. Hence a formula that λ-defines a function of natural numbers which has natural numbers as values must itself have no free symbols.

Kleene then demonstrates that the least number function

$$\epsilon \, y \, [\rho(x_1, \ldots, x_n, y) = 0]$$

is λ-definable. This function is defined to be zero if there is no such y, as in [**86**], but not as in [**85**] where it was undefined if no such y existed and where it was first shown to be λ-definable.

The final step, which had eluded others, was accomplished by Kleene due to his insight in developing the normal form theorem, Theorem IV in [**86**], where any general recursive function is shown to be expressible in the form:

$$\psi(\epsilon \, y \, [\rho(x_1, \ldots, x_n, y) = 0]).$$

Using this he proves that:[12]

> Every non-negative integral function of natural numbers which is recursive in the Herbrand-Gödel sense is λ-definable.

The second stage in the equivalence proof is to prove that all λ-definable functions. are recursive. The method used in Kleene's paper is to give a Gödel

---

[12]See [**87**, p. 347].

numbering to the λ-definition structure and so prove that the functions between Gödel numbers that correspond to functions of well-formed formulæ are recursive, and the relations between Gödel numbers that correspond to relations between well-formed formulae in λ-notation are recursive. This was a natural step for Kleene to take since he was using Gödel numbers in his parallel paper, "General Recursive Functions of Natural Numbers." and he was, by now, an expert.

In fact Kleene uses nine results on the Gödel numbering of a formal structure from Gödel's 1931 paper and a further nine results from his own paper [**86**] before he starts producing primitive recursive functions and relations pertaining specifically to λ-definability. Some of the major results he produces in the process of proving λ-definable functions recursive include the following:

i) The existence of a primitive recursive relation that is true if x is the Gödel number of a formula for a numeral $0, 1, 2, \ldots$, and the existence of the associated primitive recursive function that gives as value the number n that the Gödel number x represents.

ii) The existence of a primitive recursive relation that determines whether a number x is or is not the Gödel number of a well-formed formula.

iii) The existence of a primitive recursive relation that holds between two numbers precisely if the two numbers are Gödel numbers of well-formed formulæ that are related by the relation of immediate conversion. That is, the number x is related to the number y in this way if, and only if, x and y are Gödel numbers of two well-formed formulæ **A** and **B**, and **A** is immediately convertible into **B**. Immediate conversion does not require the transitive property of **A** conv **B**.

iv) By utilizing a key enumeration theorem, from [**86**], he proves that it is possible to enumerate, using a primitive recursive function, the Gödel numbers of all the well-formed formalæ that can be converted from a given well-formed formula **A** represented by some Gödel number x. Here the transitive property of conversion is used to generate all the well-formed formulæ produced from a given well-formed formula in as many steps as necessary.

In particular, given the Gödel number for the well-formed formula $\mathbf{L}(\mathbf{x}_1, \ldots, \mathbf{x}_n)$, then this primitive recursive function will eventually enumerate the Gödel number m for the value $\mathbf{m}$ of $\mathbf{L}(\mathbf{x}_1, \ldots, \mathbf{x}_n)$ providing such a value exists.

This enumerating function may produce the Gödel number for the value of a function many times, so the final formula used in this case specifies that the Gödel number taken is the least such number and hence the function is general recursive, rather than primitive recursive, because of this use of an unbounded least number operator.

Now if the non-negative integral function $L(x_1, \ldots, x_n)$ is λ-definable by $\mathbf{L}(\mathbf{x}_1, \ldots, \mathbf{x}_n)$ then it is known that for all $\mathbf{x}_1, \ldots, \mathbf{x}_n$ $\mathbf{L}(\mathbf{x}_1, \ldots, \mathbf{x}_n)$ will eventually convert into a formula for a numeral $\mathbf{m}$. Hence Kleene has proved that the function that produces the Gödel number of the value of $\mathbf{L}$ from the Gödel number of $\mathbf{L}$ itself is general recursive. Kleene therefore states:[13]

> Every λ-definable non-negative integral function of natural numbers is recursive in the Herbrand-Gödel sense.

Church actually announced this result first, but, as he states in a footnote to his paper, it was obtained independently by both Kleene and himself at about the same time.

6.2.2. Some Extended Equivalence Results. A significant feature of Kleene's papers is that he always covers the work very thoroughly, indeed he

---

[13]See [**87**, p. 349].

always provides a wealth of details and additional theorem beyond the immediate objective of the paper. This paper is no exception: having proved his main theorem he then extends that result and improves on the definitions and ideas contained in the paper. Thus Kleene proves that the class of λ-definable functions and recursive functions are equivalent, not only when the range of values of the independent and dependent variables are those of natural numbers, but also when the dependent variables are well-formed formulæ.

To this end Kleene first of all needs to define the term recursive for functions whose values are well-formed formulæ. While considering this we will also take the opportunity to discuss Church's definition of potential recursive functions. Although it is not particularly relevant to Kleene's work, it is relevant to an even more extended sense of recursion to which Kleene's definition points and Church also uses it later when considering effective processes.

Kleene considers a function, L, of n natural numbers $x_1, \ldots, x_n$ for which the values are well-formed formulæ, and defines L to be recursive if the function $\lambda(x_1, \ldots, x_n)$ is recursive; where $\lambda(x_1, \ldots, x_n)$ is the function which corresponds to L under the representation of formulæ by Gödel numbers. So that:

> If $L(x_1, \ldots, x_n) = \mathbf{W}$ where $\mathbf{W}$ is a well-formed formula and $\lambda(x_1, \ldots, x_n) = w$ where w is the Gödel number of $\mathbf{W}$ then L is recursive if λ is recursive.

So, to prove this wider equivalence between λ-definable functions and recursive functions where the dependent variables are well-formed formulae Kleene first needs to prove that $L(x_1, \ldots, x_n) = \mathbf{W}$ can be λ-defined by

$$\mathbf{L}(\mathbf{x}_1, \ldots, \mathbf{x}_n) \text{ conv } \mathbf{W}.$$

Church in his paper, defines the term "potentially recursive" in order to deal with functions whose independent variables are contained in a subset of the class of natural numbers. A function F is potentially recursive if it is possible to find a recursive function F' whose domain of definition is the whole of the positive integers, which agrees with the value of F for all cases where F is defined.

This definition allows Church to define the notion of recursive function for functions whose dependent and independent variables are well-formed formulae. If, as before in Kleene's definition, all well-formed formulae are replaced by Gödel numbers, then the given function will be converted into a function defined on a subset of the natural numbers since not every integer is the Gödel number of a well-formed formula. Hence any function, whose independent and dependent variables are all, or part of, the class of well-formed formulæ, is defined to be recursive if the corresponding function of Gödel numbers is potentially recursive.

Again Gödel numbering is seen to be a useful and powerful technique in that it has allowed the definition of recursive functions to be extended to a very much wider class of functions. Indeed both Kleene's and Church's papers are heavily reliant on the technique of Gödel numbering with its reduction of all sorts of formal structures to numerical functions and relations.

In order for Kleene to prove that $L(x_1, \dots, x_n) = \mathbf{W}$ can be λ-defined he first stipulates that all the values of L must have the same set of free symbols. This is because free symbols are invariant under conversion and the formulæ for the natural numbers $x_1, \dots, x_n$ contain no free symbols. Hence the values $\mathbf{W}_1, \mathbf{W}_2, \dots$ of L must contain the same free symbols irrespective of which natural numbers are used as arguments. In fact Kleene cleverly uses this fact to reduce the main part of the proof of L being λ-definable to the case where all the values of L contain no free symbols. He does this by first taking the values of L and making each free symbol into a bound symbol thus producing a new function L'. He then shows that:

L being recursive implies L' is recursive and L' being λ-definable implies L is λ-definable

which leaves him only, to prove that:

L' is recursive implies L' is λ-definable

where, of course, the values of L' contain no free symbols. This task Kleene proceeds to carry out by using Gödel numbering and primitive recursive functions and relations as before and utilizing some work on particular combinations with no free symbols which originated with Rosser. During the course of this proof Kleene introduces the λ-definition of relations by considering their representing functions in a similar manner to that of recursively defining relations.

Kleene then uses the result of this proof to show that certain sequences of well-formed formulæ are λ-enumerable, where by λ-enumeration he means that they are the value of some function that is recursive and so, by his last result, λ-definable.

Some of the more significant results include:

i) The class of well-formed formulæ having a given set of free symbols is λ-enumerable.

ii) The class of well-formed formulae with no free symbols in normal form is λ-enumerable.

iii) The class of well-formed formulæ having a given set of free symbols which have normal forms is λ-enumerable.

Kleene credits this final result to Church and it also appears in Church's [23]. It should be noted that the class of well-formed formulae which have normal forms is not recursive, and therefore λ-definable, but is only recursively enumerable, and therefore λ-enumerable. This is because, if the set was recursive, then it would be recursively enumerable in order [**86**], which would give an effective method of deciding whether a function had a normal form or not and this, as we shall see later, would contradict some important undecidability results.

The final theorem from Kleene's paper is the converse of the previous major result, namely that λ-definable functions whose values are well-formed formulæ are recursive.

Earlier in the paper, as we have already seen, Kleene has proved that all λ-definable functions whose values are natural numbers are recursive, but dealing with λ-definable functions with well-formed formulæ as values gives rise to an ambiguity.

This can be explained by considering some given well-formed formula representing the left hand side of a function of natural numbers like $\mathbf{L}(\mathbf{x}_1, \ldots, \mathbf{x}_n)$. This well-formed formula is convertible into many other well-formed formulæ $\mathbf{W}_1, \mathbf{W}_2, \ldots$ and so can be used to λ-define numerous different functions of natural numbers with well-formed formulae as values, namely:

$$L_1(x_1, \ldots, x_n) = \mathbf{W}_1$$
$$L_2(x_1, \ldots, x_n) = \mathbf{W}_2$$

and so forth. Which one should be proved recursive? The problem is that there are $2^{\aleph_0}$ functions that each $\mathbf{L}_i(\mathbf{x}_1, \ldots, \mathbf{x}_n)$ λ-defines and yet there are only $\aleph_0$ recursive functions of natural numbers. Hence not all the λ-definable functions can be recursive.

This is not the same as the case of functions with natural numbers as values, where the well-formed formulae that represent the natural numbers are unique. A similar type of condition is needed here so that a particular $\mathbf{W}_i$ can be chosen. Then Kleene can attempt the proof that

$$L(x_1, \ldots, x_n) = \mathbf{W}_i$$

is recursive starting with the observation that this formula is λ-defined by

$$\mathbf{L}(\mathbf{x}_1, \ldots, \mathbf{x}_n) \text{ conv } \mathbf{W}_i.$$

The pre-condition he uses is that the well-formed formula in question should be in principal normal form.

Principal normal form is an unambiguous variant of normal form, where the proper symbols used in the formula are in their correct order according to the given list. That is, the symbols following each λ are in the order given by the original list of proper symbols starting with the first and omitting any symbol used as a free symbol. It should be noted that the formulæ for the natural numbers $1, 2, 3, \ldots$ are in principal normal form so this condition conforms with the original proof of the theorem for functions with natural numbers as values.

Having made this restriction on the well-formed formulæ, Kleene only takes a few lines to prove that:[14]

> Every λ-definable, function of n natural numbers of which the values are well-formed formulas in principal normal form is recursive (i.e., the corresponding numerical function is recursive in the Herbrand-Gödel sense).

6.3. Church's Thesis

Church starts his 1936 paper by sketching two examples of problems in which the necessary and sufficient condition for the truth of a certain proposition in a mathematical theory is that there exists some function f that is equal to

---

[14]See [**87**, p. 353].

two if the proposition is true and is equal to one if the proposition is false. This condition would obviously be trivial unless it was also required that the function should be effectively calculable. It therefore can be seen that an essential part of the original problem was the effective calculability of the function involved. Church states the purpose of his paper thus:[15]

> ... to propose a definition of effective calculability which is thought to correspond satisfactorily to the somewhat vague intuitive notion in term is of which problems of this class are often stated, and to show, by means of an example, that not every problem of this class is solvable.

The definition he proposes in his paper is that the class of effectively calculable functions can be equated to either the class of λ-definable functions or the class of general recursive functions. This thesis has been known as Church's thesis ever since Kleene thus named it *Introduction to Metamathematics* [**91**].

The original idea for this thesis occurred when all the functions considered by Kleene and Church were found to be λ-definable. It seemed that all effectively calculable functions were λ-definable and so the original thesis, which was conceived in 1933, was in terms of λ-definable functions. After Church had spent some time worrying over it, he eventually proposed the thesis to Kleene in February or March 1934. As Kleene recalled:[16]

> He spent some months sweating over it and saying: "Don't you think it is so?" and I was a sceptic, and when he came out and asserted the thesis I said to myself: "He can't be right." So I went home and I thought I would diagonalize myself out of the class of the λ-definable functions and get another effectively calculable function that was not λ-definable. Just in one night

---

[15]See [**23**, p. 346].
[16]See [**29**, p. 7].

> I realized you could not do that, and from that point on I was a
> convert. But until Church really came out and said so, I guess
> I had not really believed they would be all of them.

A little later Gödel introduced his general recursive functions, and initiated
the idea that they might also be equivalent to the effectively calculable functions. The proof of the equivalence between the general recursive functions and
the λ-definable functions was a vital piece of evidence in support of the thesis
as finally stated in Church's paper.

The evidence that Church presented in this paper for his thesis could only
be an attempt to justify it and not to prove it since, as Church himself admitted,
it would be impossible actually to prove the equivalence of a formal definition
with an intuitive notion. In fact Church did not present the evidence for his
thesis as clearly as he might have. It appears that he was still rather hesitant
about asserting the thesis, implying that perhaps he was still agonizing over
its truth. As noted earlier, his original abstract, and therefore the first printed
statement of the thesis, only mentioned general recursive functions, whereas
his original conception had been in terms of λ-definable functions, so it seems
likely that his earlier drafts were written before the full equivalence results were
known. This may have contributed to his lack of conviction and reduced the
force with which he presented his case.

The evidence that he gives for his thesis is summarized here in four parts
of which (4) found in Section 7 of his paper represents his major evidence in
support of the thesis.

In footnote 3 on page 346 Church states:

> 1) The fact, however, that two such widely different and (in the
> opinion of the author) equally natural definitions of effective
> calcualability turn out to be equivalent adds to the strength of

the reasons adduced below for believing that they constitute as general a characterization of this notion as is consistent with the usual intuitive understanding of it.

Then on page 351 he claims that:

2) It is clear that for any recursive function of positive integers there exists an algorithm using which any required particular value of the function can be effectively calculated.

This is his claim that all recursive functions are effectively calculable functions. He proceeds to justify this claim by going through the definition of a general recursive function as formalized in [**86**] and repeated earlier in his paper itemizing the steps in the calculation and showing that each step is effective.

He discusses in footnote 10 of this page, the problem that, to some people, this definition may not be effective due to the non-constructive proof that a value of the function can ultimately be found. He suggests that the implicit existential quantifier which appears in the definition of a general recursive function can be taken in a constructive sense. In this way the result asserted in the paper, that all general recursive functions are effectively calculable, can be defended against criticism from a constructivist, since the level of constructivity required for effective calculability can be matched by the same degree of constructivity in the definition of the general recursive functions.

3) As we have already seen, on page 349 of [**23**], Church claims that, in the case of λ-definable functions, the reduction of formulæ to normal form provides an algorithm for the effective calculation of particular values of the function.

His main arguments, given in Section 7 of his paper are devoted to the converse of (2) and (3), namely justifying that all effectively calculable functions are recursive and therefore λ-definable. He does this by taking two other natural counterparts for effective calculability and showing that they can be considered to be recursive. The result then is that

> 4) No larger class of functions than recursive functions is necessary to cover all effectively calculable functions.

This result, coupled with his claim that all recursive functions are effectively calculable, will give the required equivalence. The two other possible definitions of effectively calculability that naturally suggested themselves to Church and which he introduces in the paper are:

> i) function is said to be *effectively calculable* if there exists an algorithm for the calculation of its values
> ii) a function said to be *effectively calculable* if it is calculable within a logic containing certain restrictions.

The proof that these functions are recursive is achieved by use of an inductive argument.

In case (i) the algorithm is broken down into very small steps, each of which needs to be effectively calculable for the whole to be effectively calculable. He then represents each of the steps by numerical expressions using a Gödel numbering. He can then argue that these numerical functions are sufficiently basic that they must be recursive and he then proves that this means that the whole function is recursive. In other words, if the effective calculability of each of these simple steps is taken to mean recursiveness then the whole function is recursive.

In case (ii) the aforementioned conditions on the logic used are first of all described. These include such requirements as, each rule of procedure should be an effectively calculable operation, all the rules of procedure, if infinitely many, should be effectively enumerable, all the formal axioms, if infinitely many, should be effectively enumerable and the relation between a positive integer and the expression which stands for it should be effectively determinable. These requirements are taken to mean that the corresponding Gödel number functions and relations are recursive. That is, each rule of procedure should be recursive, the complete set of rules of procedure should be recursively enumerable and so on.

He then defines a function F of one positive integer to be *calculable within the logic* if there exists an expression f in the logic such that $\{f\}(\mu) = \nu$ is a theorem when, and only when $F(m) = n$ is true, where μ and ν are the expressions which stand for the positive integers m and n. He can then prove that, with these restrictions on the logic being assumed recursive, the function F itself must be recursive.[17]

Both these proofs show that if the individual operations, rules or steps are assumed recursive then the whole function is recursive. They also suggest that the individual operations and steps are very likely to be recursive since, after Gödel numbering, they would be of a fairly simple numerical nature.

While considering the second alternative definition of effectively calculable functions it is worth noting that, in a footnote to this definition, Church credited Gödel with the essential ideas behind the recursive conditions for the logic, where the recursion used in the papers he referred to, [51] and [54], was primitive recursion. There is yet another connection with Gödel in that Church's definition of a function being calculable within the logic, given here, is essentially the same as Gödel's definition of a function being reckonable in a logic.

---

[17]He deals with the case of functions of one variable for convenience, but indicates that the results can be extended to functions of more arguments.

Gödel introduced the concept of reckonable very briefly in his 1936 paper "Über die Lähge von Beweisen" [**55**]. An English translation of this paper, "On the Lengths of Proofs," can be found in [**35**], where the notion is translated "as computable in" a logic:[18]

> ...function $\phi(x)$ may be called *computable* in S if each numeral m there corresponds a numeral n such that $\phi(m) = n$ is provable in S.

This brief exposition can be seen to describe the same functions as those given in Church 1936. Further evidence that they are meant to be the same functions can be seen by considering Rosser's review of Gödel's paper in the *Journal of Symbolic Logic* where Rosser referred to Gödel's concept as effective calculability relative to a logic.[19] He remarked that if the hypothesis that the logic was consistent was added to Gödel's definition, then effective calculability relative to the logic implied that the function was general recursive and that Church had essentially proved this in his 1936 paper.

Rosser went on to claim that the converse relationship held and therefore the concepts of general recursive functions and functions effectively calculable relative to a logic were equivalent. Gödel was certainly aware of the wide range of his functions because in a remark he added to the paper, he noted that the concept of "computable in" L was absolute and did not depend on the particular system of logic involved.[20] It seems that both Church and Gödel developed this concept independently at about the same time.

A more succinct, modern description of reckonable can be found in *Introduction to Metamathematics* where, on page 295, Kleene defines:

---

[18]See [**35**, p. 82].

[19]See [**139**, p. 116].

[20]Taken from [**35**], hence the use of the word "computable".

A number-theoretic function $\phi(x_1, \ldots, x_n)$ is *reckonable* in a formal system (or *calculable* within the system), if there is a formula $\mathbf{P}(\mathbf{x}_1, \ldots, \mathbf{x}_n, \mathbf{w})$ with no free variables other than the distinct variables $\mathbf{x}_1, \ldots, \mathbf{x}_n$ and $\mathbf{w}$ such that for each $x_1, \ldots, x_n, w$

$$\phi(x_1, \ldots, x_n) = w \ \equiv \ \vdash \mathbf{P}(\mathbf{x}_1, \ldots, \mathbf{x}_n, \mathbf{w})$$

[that is, $\phi(\mathbf{x}_1, \ldots, \mathbf{x}_n) = \mathbf{w}$ is equivalent to $\mathbf{P}(\mathbf{x}_1, \ldots, \mathbf{x}_n, \mathbf{w})$ is provable, where the variables $\mathbf{x}_1, \ldots, \mathbf{x}_n$ and $\mathbf{w}$ have been replaced by particular numerals].

This definition is very similar to the definition of a function being representable in a formal system which was given by Gödel [**54**] [21].

A function being representable in a formal system is referred to by Kleene in his book, *Introduction to Metamathematics* as a function being "numeralwise representable" in a formal system, and, in fact, on pages 295 and 296 of his book he proves the theorem that:

> If S is simply consistent, then {ϕ is general recursive} is equivalent to {ϕ is numeralwise representable in S } is equivalent to {ϕ is reckonable in S}.

Church did not present the evidence for his thesis in as convincing a form as he might have. If he had put all the evidence in one section and included a lot of extra evidence of the type that had originally convinced him of the truth of his thesis, then he would have strengthened his case. That is, he could have added comments to the effect that, all the functions that Kleene and he had tested had proved to be recursive or λ-definable and that all the methods of producing effectively calculable functions from other functions had turned out to be recursive methods.

---

[21]See Chapter 5 of this book.

Alternatively, he could have added that it was hard to imagine how an effectively calculable function could be described without deriving it by putting together simpler steps which would be almost certainly recursive. In fact part (4) of his evidence, given above, was of this nature but he did not express it very positively nor explain it very succinctly.

Since this paper any other authors have accumulated a great deal of evidence for Church's thesis. The evidence bas been summarized in several books [22] though probably Kleene's classic book, *Introduction to Metamathematics* [**91**] contains the best and most complete summary to date.

The evidence that all effectively calculable functions are recursive can be given under the four sections used by Kleene.

6.3.1. The Heuristic Evidence. This is the evidence that all the effectively calculable functions and all the methods of producing them from other functions have turned out to be recursive. It includes some function and methods specifically invented for the purpose of thoroughly testing the thesis or for deliberately trying to get a counterexample. Kleene notes that Church's formulations are, to some extent, additional examples under this heading.

6.3.2. The Equivalence of Various Diverse Formulations. In this context Church gave the two contenders that he already knew, λ-definable functions and general recursive functions, and essentially used the concept of reckonability in his evidence, as we have already seen. Kleene, however, was able to describe a few other contenders invented later as well as variants of some of the main formulations ([**91**], pp. 320-321). Over the years all of these formulations have been proved equivalent.

---

[22]See for example DeLong [**41**], Fraenkel, Bar-Hillel and Levy [**43**], Kleene [**92**], Péter [**122**], Rogers [**135**], and Schönfield [**150**].

6.3.3. Turing's Computable Functions. Turing in 1935/36 specifically set out to produce a formal counterpart to effective calculability, in contrast to the general recursive functions and λ-definable functions which had been developed before they were considered to be equivalent to the effectively calculable functions. This formulation of Turing's concept of a mechanically computable function has been singled out as perhaps the most compelling reason for believing Church's thesis. Certainly it was the reason for Gödel's final acceptance of it. This concept was proved equivalent to the class of general recursive functions by Turing himself. Details of this equivalence are explored in the next chapter.

6.3.4. Church's Formulations. The two formulations of effectively calculable functions used by Church in Section 7 of his paper.

6.4. Reaction to Church's Thesis

Church's thesis was rapidly accepted and has stood the test of time since it still has almost universal support. One or two voices of dissent have, however, been raised against the thesis. In particular Rózsa Péter has questioned whether all recursive functions are really effectively calculable and László Kalmár has argued that not all effectively calculable functions are general recursive.

Rózsa Péter's objections are of lesser importance and involve the constructive nature of the definitions. Since constructivity is considered elsewhere in this chapter no more will be said here of this objection.

The strongest criticism against Church's thesis comes from László Kalmár and his arguments can be summarized as follows. He does not believe that the class of general recursive functions is large enough to encompass all the effectively calculable functions. Instead of actually producing an effectively calculable function that is not general recursive, which would obviously directly refute Church's thesis, he attempts to show that the assumption that a particular non-recursive function is also not effectively calculable leads to some rather strange consequences.

In a lecture at the conference [**83**] Kalmár takes the function:

$$\psi(x) = \begin{cases} \mu\ y(\phi(x,y) = 0) & \exists y\ (\phi(x,y) = 0) \\ 0 & \text{otherwise} \end{cases}$$

where μ is the least number operator.

which was proved to be not general recursive in Kleene [**86**].

The function $\phi(x,y)$ used in this definition is a general recursive function and therefore, by Church's thesis, $\phi(x,y)$ is effectively calculable. Kalmár then argues that if, for some value of x, p say, there does exist a y such that $\phi(p,y) = 0$, then the effective calculability of the function $\phi(x,y)$ means that this value of y can be found and hence $\psi(p)$ can be effectively calculated.

On the other hand if we can prove by some correct methods that no y exists such that $\phi(p,y) = 0$ then $\psi(p)$ can also be calculated and equals 0. Kalmár then claims that, since ψ is not general recursive then, by Church's thesis, it can not be effectively calculated and so there must exist some value of x, q say, such that no y exists such that $\phi(q,y) = 0$ and yet there is no proof by any "correct" methods that such a y does not exist. That is, he infers the existence of a natural number q for which, on the one hand, there is no y such that $\phi(q,y) = 0$ and, on the other hand, this fact can not be proved by any correct methods.

He claims that this is rather implausible and he goes on to show that this implies that there exists an absolutely undecidable proposition, namely $\exists y\ (\phi(q,y) = 0)$, which is known to be false, and this is equally implausible.

Hence the assumption that ψ is not effectively calculable leads to some strange consequences.

Later in the paper Kalmár admits that, like Church, he can not prove that the general recursive functions plus functions like $\psi(x)$ are a larger class coexistent with the class of effectively calculable functions but that he would like to issue the proof that this is not so as a challenge. He therefore believes that ψ is effectively calculable but he does not actually go so far as to claim that he has proved this fact in his paper.

Elliott Mendelson, in a criticism of Kalmár's lecture [105], argued that the proof that no y exists such that $\phi(q, y) = 0$ must be effective for the function ψ to be effectively calculable. This means that the set of proofs by correct methods must be at least effectively enumerable. This assumption is an additional, unnoticed assumption on Kalmár part and means that all of his deductions are based on two assumptions: 1) Church's thesis and 2) that the set of proofs by correct methods is effectively enumerable. This second assumption invalidates any direct suspicion of Church's thesis since it is of itself much more likely to be false.

Mendelson also points out that Kalmár's natural number q does not refer to any particular natural number but only to same unspecified natural number, so that we can not point to any particular absolutely undecidable proposition which is known to be false.

Kalmár had not actually exhibited a function which could realistically be considered to have been proved effectively calculable and yet also be not general recursive and so his arguments provide, at best, indirect evidence against Church's thesis.

All the objectors to Church's thesis have failed to present any strong evidence against the thesis and, in particular, none of them has given any universally accepted suggestion of what an effectively calculable function, which was not also general recursive, could be like. It is not surprising, therefore, that their criticisms have gained very little support.

6.5. Recursive Unsolvability

Prior to the 1930's few mathematicians would have been willing to concede that there could be numerical problems which were easy to state but for which there existed no algorithmic solutions. This attitude can be seen in the work of Hilbert who, in 1900, said:[23]

> This conviction of the solvability of every mathematical problem is a powerful incentive to the worker. We hear within us the perpetual call: There is the problem. Seek its solution. You can find it by pure reason, for in mathematics there is no *ignorabimus*.

He repeated this in his famous lecture on the foundations of mathematics given at Münster on 4 June 1925.[24]

It was of course, partly clue to Hilbert's forceful personality and his beliefs on this subject that other mathematicians still held similar views, but his approach characterized the overall feeling on solvability of mathematical problems up to the 1930's.

---

[23]See [**69**, p. 445]. The quotation is taken from the translation of Hilbert's 1900 Paris speech by Newson [**71**].

[24]See [**76**] and [**64**, p. 384].

In 1931 Gödel's fundamental paper was published and in it he addressed himself to considering decidability in a deductive system of logic. He showed that there existed certain undecidable propositions in the logic, where, by the term undecidable, he meant that for some proposition neither it nor its negation were provable in the logic. This paper paved the way for more undecidability results and changed the attitude of mathematicians to decidability problems in general.

Hilbert, in 1917, posed the problem of solvability, in principle, of each mathematical question through finite means. If this is applied to a formal system we get the decision problem for the logic, namely, finding a decision procedure whereby, given any formula of the system, its provability or unprovability in that system can be finitely decided.[25]

Kleene, in his 1935 paper, showed that it was possible to reduce the solvability of certain, very general, elementary number theory problems to the existence of a normal form for some λ-definable formula. Church, drawing heavily on Kleene's result and similar work of his own produced at about the same time, proved the recursive unsolvability of these general elementary number theory problems and, in particular, he showed the recursive unsolvability of the decision problem for all logics subject to some very general restrictions.

Since Church had just enunciated his thesis equating recursive functions with effectively calculable functions he could claim that, providing his thesis was accepted, then he had proved the decision problem for all logics subject to same general restrictions to be effectively unsolvable. Hence Church gave the first undecidability results using effective calculation processes.

In fact Church, in this 1936 paper, addressed himself to the problem of finding effectively calculable invariants of the operation of conversion between

---

[25]The decision problem had also appeared previously in the literature in, for instance, [148] and [102].

λ-definable formulæ. He stated in a letter to Kleene of 29 November 1935 that it was he who had made the original proposal of this problem and the remark that the solution of it would imply a solution of the decision problem of *Principia Mathematica*. He continued:[26]

> ...this was made not long after our original discussion of the perpetual motion function [**p** function] and was, of course, immediately suggested by that discussion.

In his paper of 1936 he points out that there exist some obvious and fairly trivial invariants such as the number of free variables in a formula but that he would prove that a complete list of effectively calculable invariants does not exist. He argues that, should such a complete list exist, then Kleene's work in his 1935 paper would imply the solvability of most of the unsolved problems of elementary number theory. He continues:[27]

> In the light of this it is hardly surprising that the problem to find such a set of invariants should be unsolvable.

It is interesting to see that he used such an intuitive argument to justify the non-existence of the complete list of invariants since it indicates that, by this date, Church believed that most mathematicians would find such an argument compelling. In other words, by 1935 the existence of unsolvable problems was accepted by most mathematicians or at least Church felt that this was so.

This represents quite a major change of attitude from that existing prior to Gödel's theorems and prior to the work of the Princeton team. It indicates just how significant was the effect of Gödel's incompleteness theorems on the mathematicians and logicians working on foundational research at this time.

---

[26]This discussion took place around 1932, as is indicated by Church in a footnote to his 1936 paper that will be considered later.

[27]See [**23**, p. 358].

The results derived by Church in this section of his paper are first:[28]

> Lemma. The problem, to find a recursive function of two for-
> mulas **A** and **B** is 2 or 1 according as **A** conv **B** or not, is
> equivalent to the problem, to find a recursive function of one
> formula **C** whose value is 2 or 1 according as **C** has a normal
> form or not.

He means by this that the recursive solvability of the statement "**A** conv **B**" is equivalent to the recursive solvability of the statement "**C** has a normal form."

Church proves this result by utilizing Gödel numbers. He generates all the formulæ convertible from **a**, the formula for the Gödel number of **A**, and compares them all to the formulae convertible from **b**, the formula for the Gödel number of **B**. He is aided considerably in this proof by the use of results and functions from Kleene's 1935 paper, in particular Church uses Kleene's **p** function.

Church acknowledges, in the footnote that we have mentioned previously, that around 1932, when Kleene was discussing with him the properties of this **p** function, Kleene proposed the essential parts of this lemma, although the full result on the complete set of invariants was proposed by Church. That is, Kleene asked whether it was possible to find an effective method of determining for any two formulae **A** and **B** whether **A** conv **B** or not and whether it was possible to find an effective method of determining for any formula **C** whether it had a normal form or not.

Of course after Church's thesis had been conceded then this effective method meant a recursive method and so the lemma was stated in terms of recursive functions. It should be noted that the recursive function involved had

---

[28]See [**23**, p. 359].

well-defined formulæ as arguments and so depended on the previously given definition of potential recursive functions.

After the lemma Church's major result is stated and proved:[29]

> Theorem XVIII. There is no recursive function of a formula **C**, whose value is 2 or 1 according as **C** has a normal form or not. That is, the property of a well-formed formula, that it has a normal form is not recursive.

This, together with the lemma, immediately gives:[30]

> Theorem XIX. There is no recursive function of two formulas **A** and **B**, whose value is 2 or 1 according as **A** conv **B** or not.

This final theorem is the formal expression of the statement discussed at the beginning; viz. there is no complete set of effectively calculable invariants for conversion. Again Church's thesis means that the theorem can be stated in terms of the existence of a suitable recursive function.

Theorem XVIII is proved by contradiction. A recursive function, H, of one positive integer with the property that:

$$H(m) = \begin{cases} 2 & m \text{ is the Gödel number of a formula} \\ & \text{which has a normal form} \\ 1 & \text{otherwise} \end{cases}$$

is assumed to exist.

Now, using a previous theorem which states that there is an effective (recursive) enumeration of formula which have a normal form, the sequence $A_1, A_2, \ldots$

---

[29]See [**23**, p. 360].
[30]See [**23**, p. 363].

is defined to be such as enumeration of these formulæ. For each $\mathbf{A}_n$, $\{\mathbf{A}_n\}(\mathbf{n})$ is formed

Using the function H it is possible to determine effectively (recursively) of a formula $\{\mathbf{A}_n\}(\mathbf{n})$ whether it has a normal form, and if so, whether it has a principal normal form which is a formula for a numeral $1, 2, 3, \ldots$..

Now a function E is defined by:

$$E(n) = \begin{cases} 1 & \{\mathbf{A}_n\}(\mathbf{n}) \text{ is not convertible into a formula} \\ & \text{for a numeral} \\ m+1 & \{\mathbf{A}_n\}(\mathbf{n}) \text{ conv } \mathbf{m} \text{ and } \mathbf{m} \text{ is the formula for a numeral.} \end{cases}$$

As previously stated, E is effectively calculable due to the existence of H and so E is a recursive function and can therefore be λ-defined by a formula $\mathfrak{e}$. That is, $\mathfrak{e}(\mathbf{n})$ conv $\mathbf{m}+\mathbf{1}$ if and only if $\{\mathbf{A}_n\}(\mathbf{n})$ conv $\mathbf{m}$, otherwise $\mathfrak{e}(\mathbf{n})$ conv $\mathbf{1}$.

From this it can be seen that, even though $\mathbf{e}$ has a normal form, there is no n such that:

$$\mathfrak{e}(\mathbf{n}) = \{\mathbf{A}_n\}(\mathbf{n})$$

that is

$$\mathfrak{e} \neq \mathbf{A}_n$$

for any n.

So $\mathfrak{e}$ has a normal form and yet is not equal to any of the sequence of formulæ which have a normal form. This contradiction proves that the function H can not exist.

After Theorem XIX Church could prove his result on the unsolvability of the decision problem for systems of logic.

Actually Church proves that there is no effective algorithm or recursive method with which to test a formula of the logic to determine whether or not it can be derived from the rules or proved in the system. This decision problem, in terms of provability, was referred to by Church as the first form of the decision problem. He was to consider the second form of the decision problem, in terms of validity, in his paper "A Note on the Entscheidungsproblem" [22].

Church starts by characterizing the logics to which his theorem would apply, namely ones that are ω-consistent and strong enough to allow certain comparatively simple methods of definition and proof. That is, logics that are sufficient to define elementary number theory within them. In particular the system of *Principia Mathematica*, if ω-consistent, would come under this specification.

He then shows that such logics would contain a proposition $\psi(a, b)$ of two arguments a and b, such that if a and b are the Gödel numbers of formulæ **A** and **B** then $\psi(a, b)$ expresses the fact that **A** conv **B**. He then proves that if **A** conv **B** then $\psi(a, b)$ is provable in the system and if **A** is not convertible into **B** then the ω-consistency of the logic guarantees that $\psi(a, b)$ is not provable in the system.

Now suppose we can effectively determine whether or not each formula in the system of logic is provable, in particular whether or not $\psi(a, b)$ is provable for arbitrary a and b. Then we would effectively know whether or not **A** conv **B** for any **A** and **B**. This would contradict Theorem XIX so we must conclude that the decision problem for the system of logic is not effectively solvable.

Church had therefore produced the first unsolvability result for a decision problem. Prior to this there had been a few positive results for decision problems, such as Post's truth table method for the propositional calculus and Presburger's decision procedure for a formal system without multiplication, but no unsolvability results. The reason for there being no unsolvability results was that, to prove that there did not exist an algorithm for a class of problems required a precise understanding of what an algorithm is.

A characterization of all possible algorithms would be needed and this was not available prior to the 1930's. Church's thesis supplied that precise characterization by equating all algorithms, or all effectively calculable functions, with the class of general recursive functions. Thus Church's thesis was a necessary prerequisite for any unsolvability result for a decision problem.

The logic specified in Church's 1936 paper was any system of logic which was (1) ω-consistent and (2) strong enough to allow certain comparatively simple methods of definition and proof. Rosser, in "Extensions of Some Theorems of Gödel and Church" [**138**], removed the restriction of ω-consistency from Church's undecidability result, as well as from both Gödel's theorems, and replaced it with the simpler assumption of consistency.

Before this result of Rosser's, though, Church himself extended his undecidability result to the whole of Hilbert and Ackermann's *engere Funktionenkül* (the first order function calculus or predicate calculus).[31]

Church proved this extended result by first augmenting Hilbert and Ackermann's system to another system L. This system L was certainly adequate for defining simple arithmetic and, in the literature, had already been proven consistent.

To be able to apply the undecidability theorem from his 1936 paper Church also needed L to be ω-consistent. But instead of w-consistency Bernays proved, in his lectures at Princeton in 1936, that the system L possessed a weaker but sufficient property so that Church's undecidability theorem could still be applied. That property was:[32]

---

[31]This was proven in "A Note on the Entscheidungsproblem" [**22**].

[32]The result quoted here is the corrected version given in [**35**, p. 112]. The correction appeared in the same volume of the *Journal of Symbolic Logic* as did the original paper by Church

If P contains no quantifiers and $(\exists \mathbf{x})P$ is provable in L then some one of $P_1, P_2, P_3, \ldots$ is provable in L (where $P_1, P_2, P_3, \ldots$ are respectively the result of substituting for $\mathbf{x}$ the symbols for $1, 2, 3, \ldots$ throughout P).

Armed with this property for the system L, Church could invoke his undecidability result from his 1936 paper and deduce that the decision problem for the system L was unsolvable.

Having dealt with the system L Church proceeded to show that if the decision problem for Hilbert and Ackermann's *engere Funktionenkalkül* was solvable then the decision problem for the system L would also be solvable. Hence, by contradiction, Church had proved that the decision problem for the *engere Funktionenkalkül* was unsolvable.

6.6. Constructivity and Conclusion

Hilbert had claimed that the problem of effectively determining whether or not any formula of his *engere Funktionenkalkül* was valid was the fundamental problem of mathematical logic. Church in [**22**] had proved that the deducibility problem for Hilbert's *engere Funktionenkalkül* was unsolvable. That is, the problem of effectively determining whether or not a formula from the system was provable from the axioms was unsolvable.

Church continued in this paper by noting that the second form of the decision problem, in terms of validity, would similarly be unsolvable if one invoked Theorem I from Gödel's 1930 paper, which says:[33]

> Every valid [universally valid] formula of the restricted functional calculus [Hilbert's *engere Funktionenkalkül*] is provable.

---

[33]See [**64**, p. 584].

Having stated this fact Church proceeded to consider the constructive nature of his proofs. He asserted that the proof that the first form of the decision problem (the deducibility problem) was unsolvable was constructive, but the proof that the second form of the decision problem was unsolvable was non-constructive since it depended on the non- constructively proved theorem form Gödel 1930. He finished:[34]

> The unsolvability of this second form of the Entscheidungsproblem [decision problem] of the *engere Funktionenkalkül* cannot, therefore, be regarded as established beyond question.

This statement says more about Church's own doubts on the matter and his general concern about constructive proofs and methods than about the probability of the proof being acceptable to the great majority of mathematicians. As Davis says, in his introduction to Church's paper in *The Undecidable*:[35]

> Most mathematicians would probably have greater confidence
> in non-constructive mathematics than that expressed here.

In fact there are several statements of a constructivist nature in both Church's work and Kleene's papers. This seems in part due to the presence of Bernays who was Visiting Professor in Princeton during this period. Bernays seems to have read and made helpful comments on many of the earlier versions of the papers from Church, Kleene and Rosser and then reviewed quite a few of the final papers after they had been printed. Some of these comments were of a constructivist nature. For example, Church admitted, in the correction of [22], that the idea of distinguishing between constructive and non-constructive

---

[34]See [35, p. 115].
[35]See [35, p. 109].

proofs for the decision problem of Hilbert and Ackermann's *engere Funktio-nenkalkül* was, in fact, due to Bernays and so the discussions on this topic are not to be found in the original paper but in the correction [21]. That is, the constructivist comments were only made after Bernays' intervention.

Church added a footnote pertaining to constructivity when be gave the definition of a recursive function in his 1936 paper and claimed that the definition gave an effective algorithm for the calculation of its values. That note was to the effect that the existential quantifier contained in the definition of a recursive function should be understood to be of the same level of constructivity as that required of the existential quantifier in the definition of an effectively calculable function.

In this way Church hoped that the converse of his thesis that all general recursive functions are effectively calculable could be defended against any constructivist. However, it was criticisms of this constructivist nature that were behind Rózsa Péter's objections to Church's thesis. The objections voiced by Wang were also in a similar vein, as can be seen from the following quotation from his book, *Logic, Computers and Sets*.[36]

> It now seems clear that the notion of recursiveness can at most
> be considered a satisfactory substitute of the intuitive notion
> of effective procedures in a classical sense, not in a strictly
> constructive sense.

This viewpoint is by no means widely held today and most mathematicians accept Church's thesis with fewer reservations.

---

[36]See [166, p. 89].

Another example of a non-constructive proof, which was again pointed out by Bernays, can be found in Church's [**23**] where he proves a corollary to his theorem XVIII, namely:[37]

> The set of well-formed formulas which have no normal form is not recursively enumerable.

The proof of this corollary proceeds by contradiction. If it is assumed that the set of well-formed formulæ which have no normal form is recursively enumerable then, together with the previously proved fact that the set of well-formed formulæ which do have a normal form is recursively enumerable, we would possess an effective method of telling of any well-formed formula whether or not it has a normal form. This would be accomplished searching through both enumerations until the well-formed formula in question is found. But this directly contradicts Theorem XVIII.

Following Bernays' observation, Church indicates in a footnote how the proof can be converted into a constructive proof.

Church follows this with a comment on the corollary to the effect that it shows an example of an effectively enumerable set, the set of well-formed fomulæ, which is divided into two non-overlapping subsets, one of which is effectively enumerable and the other not. Indeed he continues by asking whether the second set could be more accurately considered as non-enumerable rather than non-effectively enumerable since he contends that it is difficult to attach any sensible meaning to an enumerable but not effectively enumerable set.

---

[37]See [**23**, p. 362].

Church then gives a second corollary where he describes a nonrecursive function:

$$F(n) = \begin{cases} 2 & n \text{ is the Gödel number of a formula} \\ & \text{which has a normal form} \\ 1 & \text{otherwise.} \end{cases}$$

The fact that F is non-recursive follows immediately from Theorem XVIII, but Church continues by asking whether the function F can be considered to exist at all.

He reasons that for the sequence of positive integers $F(l), F(2), F(3), \ldots$ there is no effective method by which, given any n, the nth term can be calculated. But he argues that it is also impossible to select a particular term of the sequence, the rth term say, and prove that its value can not be calculated. So it can be said that even though there is no effective method of calculating all the terms of the sequence, each of the terms individually can be calculated. It would seem that extensions of arguments of this nature were probably what led Kalmár to his criticisms of Church's thesis since there is a certain amount of similarity between Church's comments and Kalmár's criticisms.

This chapter has completed the description of the major research work carried out by Church, Kleene and Rosser in Princeton in the first half of the decade starting in 1930. We have seen the development of two main candidates for the formal equivalence to the class of effectively calculable functions and we have seen these functions put to use in proving various theorems.

Of outstanding importance amongst these theorems is the result, proved in Church's 1936 paper, that the decision problem for Hilbert's predicate logic was unsolvable. This result was the first of a whole string of further results on unsolvability.

Church, in a postcard to Kleene dated 19 May 1936, expressed the hope that decision problems would be proved unsolvable in other branches of mathematics not specifically related to logic. His hopes have since been fulfilled in work initiated by Post [130] and Markov [103].

Unsolvability results have been established for decision problems in branches of mathematics as diverse as algebra, topology and real-variable analysis. In the next, and final, chapter we consider another line of development for finding a formal counterpart to the class of effectively calculable functions, namely Turing's work on mechanically computable functions.

CHAPTER 7

Turing's Computable Functions

7.1. Introduction

The development of both the λ-definable functions and the general recursive functions was largely completed before they were considered as formal counterparts for the class of effectively calculable functions. In this chapter we give a very brief introduction to another class of functions that was created by Alan Turing specifically for the purpose of characterizing mathematically all effective calculation procedures. It is only a light sketch of Turing's work. It is included because Turing's analysis of the concept of a mechanical procedure is generally regarded as providing the most compelling reason for believing Church's thesis. Gödel, in particular, considered it most important in that it furnished a precise and adequate definition of the notion of a formal system.

Alan Turing completed his degree in mathematics at Cambridge in 1934 and proceeded to undertake some research work. He had a practical interest in computing and he even started to build a machine for computing the Riemann zeta function, cutting the gears for it himself. In 1935 his interest in computing led him to consider just what sort of processes could be carried out by a machine. This interest, coupled with his knowledge of Hilbert's decision problem for logic, to which he was introduced by his tutor, M. H. A. Newman, led him to consider

all effective calculation procedures. As Newman states in his biography of
Turing:[1]

> The Hilbert decision-program of the 1920's and 30's had for its
> objective the discovery of a general process, applicable to any
> mathematical theorem expressed in fully symbolical form, for
> deciding the truth or falsehood of the theorem. A first blow
> was dealt at the prospects of finding this new philosopher's
> stone by Gödel's incompleteness theorem (1931) which made
> it clear that truth or falsehood of A could not be equated to
> provability of A or not-A in any finitely based logic, chosen
> once for all; but there still remained in principle the possibility
> of finding a mechanical process for deciding whether A, or not-
> A, or neither, was formally provable in a given system. Many
> were convinced that no such process was possible, but Turing
> set out to demonstrate the impossibility rigorously.

According to his mother, Sara Turing, he was always very good at combining theoretical work with practical experiments. Not surprisingly, therefore,
while Turing was considering these effective calculation procedures – calculations that could be done by a mechanical process – it occurred to him that this
ought to mean that the calculations could actually be performed on a machine.
He then set about analyzing the general notion of a computing machine which
he described in his fundamental paper, "On Computable Numbers, with an Application to the Entscheidungsproblem" [**160**]. Turing subsequently reported to
Gandy that the idea for the whole of this paper came to him all at once while
he was relaxing on Grantchester meadows, Cambridge, in the summer of 1935[2].

---

[1]See [**113**, p. 256].

[2]This was communicated to the author when in discussions with Robin Gandy in Oxford
on Monday 21st March 1977

The machines that Turing described in his paper almost immediately became known as Turing machines. The first reference to them under this name occurred in Church's review of Turing's 1936 paper [**24**]. This paper was dated 25th May 1937. Subsequently all similar machines have been referred to as Turing machines.

In his paper Turing not only described the machines that would compute numbers but also considered many wider issues. Included amongst these was the existence of a universal machine that could be programmed to do the task of any particular machine. This was a forerunner of our general purpose computers.

Turing also considered in some detail the types of real numbers that could be computed by one of his machines and the evidence for his thesis which equated these numbers with the numbers that could naturally be regarded as computable. Finally, Turing adapted his machine to produce provable formulæ for the Hilbert and Ackermann functional calculus and thereby proved that the decision problem for this calculus was unsolvable.

Prior to Turing's work Church bad already formulated his thesis equating λ-definable functions, or recursive functions, with effectively calculable functions. Church's 1936 paper, containing this thesis, came out in the first half of 1936. Turing's work, though, was done without any knowledge of the work of Church, Kleene or Rosser that was carried out in Princeton. As Turing was about to send his paper to press in May 1936 he received an abstract of Church's paper and immediately sent for a copy.[3] After reading Church's paper Turing added an appendix to his original paper, dated 28 August 1936, containing a brief proof of the equivalence of his computability with Church's λ-definability. Turing followed this up with a further paper [**161**] in which he gave a more

_____

[3]Turing's paper was received for publication by the London Mathematical Society on 28 May 1936.

complete proof of the equivalence between his computable functions, the λ-definable functions and the general recursive functions.[4]

Significantly, another paper appeared in 1936 written by Emil Post [**128**]. This paper was received for publication on 7 October 1936. Written independently of Turing's paper but not independently of the work at Princeton, this paper by Post contained a formulation remarkably similar in nature to Turing's and subsequent descriptions of Turing machines have contained some factors from both Turing's and Post's original papers.

7.2. Turing's Machines

Turing introduces his fundamental paper with the following comment:[5]

> The "computable" numbers may be described briefly as the real numbers whose expressions as a decimal are calculable by finite means. Although the subject of this paper is ostensibly the computable *numbers,* it is almost equally easy to define and investigate computable functions of an integral variable or a real or computable variable, computable predicates, and so forth.

Turing chose to deal mainly with computable numbers in his paper because be considered them to be less cumbersome, but computable functions are fundamentally more useful and fitted in with the work at Princeton, in that they could be equated with the effectively calculable functions, so almost all mathematicians have subsequently used his ideas and techniques to define the class

---

[4]Having completed his original paper Turing was invited to Princeton to work with Church. Although he addressed his appendix from Princeton, 28 August 1936, he did not actually arrive there until late September, having set sail for New York on 23 September 1936.

[5]See [**160**, p. 230].

of computable functions. Turing, himself, actually defined and considered computable functions later in his paper and also in his next paper, "Computability and λ-Definability," [161] but it should be remembered that his main definitions and the details of the 1936 paper were for sequences that represented the digits of a number.

Following his informal definition of a computable number, given in the above quotation, he proposes the formal definition of a computable number as a number whose decimal can be written down by a machine. This definition obviously requires some amplification, in that he first needs to explain what he means by a machine, what it is to consist of and what it is allowed to do. He does this by imagining an idealized computing machine. This is a machine that is supplied with a one-way potentially infinite tape divided into squares. At any moment the machine can scan just one square and read the symbol, if any, on it. The machine is allowed to print one of a finite number of symbols on the tape if the square is blank, or erase a symbol, or move one place to the left ar to the right.

The tape is to be subdivided into two alternate sequences of squares which he calls E-squares and F-squares. The E-squares are for rough working and any symbol placed on an E-square is eligible for erasure. The F-squares are to contain the continuous sequence of digits that will form the decimal expansion of the number that is being computed. Any symbol on these squares can not be erased.

In order that the machine can store previous actions it is capable of taking up one of a finite number of different conditions $q_1, \ldots, q_R$ called m-configurations. The behavior of the machine at any instant is to be determined by the symbol, S_j, on the square it is scanning together with the m-configuration that indicates the condition of the machine at that moment. After printing, erasing or moving to the left or right the m-configuration can be changed. The two pieces of information that can determine the action of the machine at any instant, the m-configuration and the scanned symbol, are called the configuration of the machine.

A machine whose action at any point is completely determined by its possible configurations is called an automatic machine. Turing only considers in detail automatic machines in this paper but he does indicate that it is possible to imagine another type of machine, called a choice machine. A choice machine is a machine that, at certain points in its operation, requires some external stimulus. It will stop at these points and its future action will then be dependent on the external information supplied.

The complete behavior of an automatic machine can therefore be described by giving a table consisting of possible configurations at the start of each act, the behavior to be enacted for each of these configurations and the m-configuration that it is to take up after the action. The information given in such a table completely determines the machine in question.

Turing specifies that the symbols that can be printed on the F-squares are 0's and 1's only. The decimal expansion of the number being computed is represented by a sequence of 0's and 1's. Each digit is the number of 1's between successive 0's. Hence the real number 0.24531 ... would be represented by the sequence:

$$0110111101111101110100...$$

Thus Turing's original machines were designed to continuously print the 0's and 1's that represented the decimal expansion of some real number between 0 and 1.

If an automatic machine, described by some table of moves, is supplied with a tape that is either blank or has some specified set of symbols on it, then it can take one of the following actions:

i) continue for ever printing 0's and 1's on the F-squares,

ii) reach a point where there is no possible move as specified by the table

iii) reach a point where it only prints symbols on the E-squares and not the F-squares.

In case (i) it is called a circle-free machine, in cases (ii) and (iii) it is called a circular machine.

Hence, in order to compute a sequence, and therefore a number, a machine has to be circle-free. Turing sets the stage:[6]

> A number is computable if it differs by an integer from the number computed by a circle-free machine.

A number is defined to be computable therefore if there exists a computing machine which has a finite set of symbols and a potentially infinite tape on which to print, and which can write down the sequence of its digits to any desired length if allowed to run for sufficient time.

Later in the paper Turing defines computable functions. One of the methods he uses to define a computable function of an integral variable is a function $\xi(\gamma, n)$ of arguments γ, a computable sequence in which 0 appears infinitely often, and n, a positive integer, is defined as

$\xi(\gamma, n) \overset{\text{def}}{=}$ the number of 1's between the nth and the $(n+1)^{st}$ 0 in γ.

The function $\phi(n)$ can then be defined to be computable if, for all n, there is a γ such that

$$\phi(n) = \xi(\gamma, n).$$

_____

[6]See [**160**, p. 233].

Hence, the successor function $S(n)$, defined for a positive integer n is computable if there exists a computable sequence γ such that

$$\gamma = 0110111011110\ldots$$

Turing proceeds to show how it is possible to build up quite complicated sequences from these very simple beginnings. He illustrates some tables for machines that are capable of performing certain standard processes, like copying a sequence of symbols, or comparing sequences, or erasing symbols of a certain kind. These operations can be used many times when producing a more complicated sequence.

To facilitate this use he defines, for each such process, a function of the configurations involved, so that each process can be used for many different m-configurations and symbols. These m-functions, as they are called, are only introduced as abbreviations but they show in a practical form how it is possible for much more complicated operations and sequences to be built up.

In fact, the examples of computable sequences and computable operations illustrated by Turing were really designed to be part of the table for a universal machine. Consequently some of the operations described by Turing were intended for a tape that already had something printed on it; namely, the partial workings of the universal machine. Unfortunately Turing often failed to specify what was supposed to be on the tape at the start and, on first reading his paper, one may very easily think that Turing's descriptions are wrong and that his operations do not achieve what he says they do. Not only this, Turing did actually make quite a few minor errors and Post included his paper "Recursive Unsolvability of a Problem of Thue" [**130**] an appendix itemizing some of Turing's errors and giving their corrections. In this paper Post also gave an updated description of a Turing machine.

Turing's paper thus shows some signs of haste, as though his own thought processes were so far ahead of what he was actually writing that be failed to give the reader sufficient information to be able to follow all the details exactly. But it is very clear that the overall processes Turing had in mind would work. The essential ideas of the paper come through very distinctly and the fundamental importance of this paper is not lost even though a few details are missing.

Turing was not solely interested in describing the workings of machines for calculating sequences but had in mind much wider issues such as general questions of computability and the existence of a universal machine.

His first step in describing his universal machine is to illustrate how the operations of each specific machine can be numbered. Turing shows that by introducing more m-configurations it is possible to reduce each line of the table for a machine to one of three standard forms:

$$N_1 : q_i S_j S_k L q_m$$
$$N_2 : q_i S_j S_k R q_m$$
$$N_3 : q_i S_j S_k N q_m$$

where

    i) q_i is the m-configuration at the start of the step

    ii) S_j is the symbol scanned

    iii) L, R and N represent move left, move right and no move respectively

    iv) S_k and q_m are the scanned symbol and m-configuration at the end of the step.

The whole table for a machine can then be described by writing down each of these lines separated by semi-colons. This list will then represent a complete description of the machine in that it represents a list of all the machine's instructions.

The standard description for the machine is obtained from this by:

    i) replacing each q_i by the letter D followed by the letter A i times

    ii) replacing each S_j by the letter D followed by the letter C j times

Thus $q_1 S_0 S_1 R q_2$ has the standard description:

$$D\ A\ D\ D\ C\ R\ D\ A\ A.$$

Finally the description number for the machine is produced by replacing the letters A, C, D, L, R, N and the semi-colon by the numbers $1, 2, \ldots, 7$. Thus $q_1 S_0 S_1 R q_2$ now has description number

$$3\ 1\ 3\ 3\ 3\ 2\ 5\ 3\ 1\ 1.$$

The description number for a machine is not unique since redundant lines can always be added to the original table. A number which is a description number of a circle-free machine is called a satisfactory number.

The idea of reducing formal structures and operations to numbers was by now well established following Gödel's first use of such a technique in 1931. Turing's numbering, though, was quite different in structure from Gödel's in that he just produced a long line of digits whilst Gödel would number each sub-part and use a product of primes to combine these parts. This difference was because Turing was only interested in producing a number that could be read by a machine, while Gödel numbering was designed to allow primitive recursive functions and relations between the numbers to represent specific operations on the original structure.

Turing can now describe a universal machine. It is a single machine, $\mathcal{U}$, that, when supplied with a tape containing the standard description or description number of same machine $\mathcal{M}$, will compute the same sequence of digits that the machine $\mathcal{M}$ would compute. Thus one machine can operate like any other machine providing it is first supplied with the coded description of that machine. Turing shows that such a universal machine $\mathscr{U}$ can exist by actually giving a table for $\mathscr{U}$ in terms of the abbreviations, the m-functions, introduced previously.

Before discussing the wider issues associated with Turing machines that were dealt with later in Turing's original paper, we shall consider Post's version of a mechanical computing device and some more modern formulations.

7.3. Post's Formulation and Later Devices

Post conceived of his machine quite independently of Turing. Post in his 1936 paper [128] did not actually refer to his formulation as a machine but considered an idealized worker or human operator moving in a certain symbol space according to some given set of directions.

Turing imagined a machine with a one-way infinite tape which was allowed the use of n symbols for computing unending sequences of 0's and 1's. Post, on the other hand, imagined a mechanical worker supplied with a two-way infinite sequence of spaces or boxes and the ability to mark the space or box that be was in with just one symbol, like a vertical stroke.

Post assumed that his mechanical operator was capable of performing essentially the same acts as Turing had assumed for his machine, namely moving left, moving right, marking a box, erasing the mark in a box or reading the contents of a box. As already stated, Post imagined his worker performing these acts according to a given set of directions. These directions would be a list of instructions informing the worker what to do at each stage and in which direction to go next.

Post did not specify how these directions would work in practice and, indeed, he never had to since Turing had worked out all the details for his machines and it was obvious that Turing's formulation allowed his machines to perform all the actions that Post had in mind for his worker. Consequently all subsequent descriptions of similar machines have used Turing's internal states technique as the basis for their detailed operations.

Post's mechanical worker was designed to perform certain symbolic manipulations on data that were coded onto the symbol space by some outside agency. The directions supplied to the worker were intended to produce solutions, if such solutions existed, to some given general problem consisting of a whole class of specific problems. The data given symbolically by the outside agency was to represent one of these specific problems and the worker had to react to the symbols and operate on them according to the given set of directions. If and when he had produced the solution to the problem he would stop. For different initial information supplied by the outside agency, representing a different specific problem, the set of directions would allow the worker to produce a different solution for interpretation by the outside agency.

Turing, on the other hand, specified that a blank tape should be supplied to his machine and consequently the sequence that the machine produced was completely predetermined by the table of configurations for that machine. In this respect, therefore, Post's operator was more like Turing's universal machine which did need an initial tape configuration on which to work and would then produce the sequence required by this initial information. Turing's formulation, of course, did not include a stop instruction and his machines continued for ever except circular machines, which could stop. But then they didn't successfully compute a sequence either.

Almost all modern descriptions of Turing machines envisage the machine being used to compute the value of some function. The argument values would be supplied in symbolic form on the tape at the start and the machine would compute the value of the function, if it could, and then stop.

Such modern descriptions of Turing machines utilize ideas from both the original papers, [160] and [128], as well as from an updated version given in [130], "Recursive Unsolvability of a Problem of Thue." In Turing's paper, as already noted, the actions of a machine at each instant were given by a quintuple such as

$$q_i \ S_j \ S_k \ R \ q_m$$

This method is still used sometimes, but more frequently the formulation given in [130] is found preferable. Turing's description required the machine to print or overprint a symbol for each act. In Post's new format no printing was required when the machine moved left or right. That is, Post described each act using a quadruple, for example

$$q_i \ S_j \ L \ q_l \ ; \ q_i \ S_j \ R \ q_l \ ; \ q_i \ S_j \ S_k \ q_l$$

A machine whose table was described in this manner could then be converted into a choice machine by just adding one extra quadruple

$$q_i \ S_j \ q_k \ q_l$$

which would instruct the machine to go to state q_k or q_l depending on the nature of the external information supplied.[7]

So the basic operations of all modern machines follow the method first expounded by Turing in his original paper. The convention, introduced by Turing in this paper, of utilizing alternate sequences of squares for the main sequence under construction and for rough working, coupled with certain restrictions on their use has not, however, been repeated. Post in his 1947 paper, described these machines as Turing-convention machines and illustrated that this convention was not necessary to the operation of machines.

---

[7]This is the description of Post's machine given in Davis' book, *Computability and Unsolvability* [34].

7.4. Turing's Thesis

Church [23] claimed that all effectively calculable functions were λ-definable or general recursive and this has since become known as Church's thesis. In a similar manner Turing introduced an equivalent type of thesis into his paper. Turing's thesis proposed that all the numbers which would naturally be regarded as computable could also be computed using one of his machines. Turing recognized that any arguments he gave in support of his thesis must necessarily be appeals to intuition and that he could not prove his result mathematically.

In his paper he considers three kinds of evidence:

a) direct appeal to intuition.

b) a proof that numbers computable by machine are equivalent to numbers computable using Hilbert and Ackermann's functional calculus, in case this new formal equivalent to "numbers naturally regarded as computable" is preferred.

c) he gives copious examples of classes of numbers which can be computed by machine.

The arguments given in section (a) involve Turing in analyzing all the effective processes that a human computer would carry out when computing a number according to sane set of explicit instructions, and showing that these can all be performed on some sort of machine.

Turing recognizes that once he has established this relationship then he will also have established other propositions of a similar nature. That is, Turing machines can be used equally well for other symbolic manipulations, such as calculating function values or generating all the provable formulæ for the Hilbert and Ackermann functional calculus. Hence, Turing's analysis of the effective processes that a human computer would carry out when computing a number, according to certain fixed procedures, would also apply to a human computer calculating a function value or attempting to show that a formula of the Hilbert and Ackermann functional calculus was provable, if this was possible. Turing is claiming that, in each case, if there is a general procedure for a human computer to do these things, then a Turing machine can also carry out that task.

Turing justifies his claim that computations done by a human computer can be done on a machine by analyzing the limitations under which a human computer would work. These limitations include the supposition that the calculation is done on a two dimensional piece of paper. Turing asserts that it is fairly obvious that the two dimensional character of the paper is not essential to the calculation, and that therefore the calculation can also be done on a one dimensional array. He assumes that the number of symbols used by a human computer must be limited otherwise:[8]

> If we were to allow an infinity of symbols, then there would be
> symbols differing to an arbitrarily small extent.

The behavior of the human computer at any instant is determined by the symbol under observation and his state of mind, where the number of symbols observed at any one time is bounded and the number of states of mind is finite otherwise:[9]

> If we admitted an infinity of states of mind, same of them will
> be "arbitrarily close" and will be confused.

Any operation performed by a human computer can always be split up into simpler operations. Each simple operation is then of the form (1) change just one symbol on an observed square, or (2) change just one observed square for another square, along with a possible change of state of mind in each case. In

---

[8]See [**160**, p. 249].
[9]See [**160**, p. 250].

the second of these simple operations the new observed square must be within
some fixed number of squares from the original set of observed squares.

Also it is possible to include amongst the observed squares those that are
marked in some way and therefore immediately recognizable. But again the
number of these marked squares can not be too large or else they could hardly
be claimed to be easily identified. As an example Turing considers the case of
numbering theorems and claims that if the numbers used rise too high, then
the theorems in question can not be easily distinguished.

Turing argues that all the limitations under which a human computer works
are such that it is obvious that a machine, not much different from his previously defined ones, can do the computation, and hence the calculation must be
reducible to one that can be done on a Turing machine.

Essentially Turing's machines were defined, in the first place, to simulate
the operation of a human computer and consequently this argument parallels
his original thought processes when he designed the machines. It is therefore
not surprising that Turing considered that his arguments proved that anything
a human computer could effectively calculate could also be done on a machine.

Under part (b) of his evidence, Turing considers the Hilbert and Ackermann. functional calculus. He claims that he has the description of a machine,
$\mathcal{K}$, that will generate all the provable formulæ of this logic. He then considers
a sequence α and defines the proposition $G_\alpha(x)$,

$$G_\alpha(x) \stackrel{\text{def}}{=} \text{The } x^{th} \text{ figure of } \alpha \text{ is } 1.$$

He shows that it is possible to construct a formula $\mathfrak{U}$ of the Hilbert and Ackermann functional calculus, in terms of $G_\alpha(x)$ and other axioms and functions
of the logic that defines α. In these circumstances α is said to be computable,

that is, if there is a formula $\mathfrak{U}$ of the Hilbert and Ackermann functional calculus that defines a then α is computable.

With this alternate definition of a computable sequence Turing proceeds to show that if α is computable using the Hilbert and Ackermann functional calculus then there exists a machine $\mathcal{K}_\alpha$, derived frcm the machine $\mathcal{K}$, that will compute α and hence α is computable by machine. Also, by describing machines in terms of the functional calculus, Turing asserts that if α is computable by machine then it is computable using the functional calculus.

Consequently we see that part (b) of Turing's evidence represents a proof of the equivalence of the definitions "α computable by machine" and "α computable by functional calculus."

The final evidence given by Turing, under part (c), is in the form of numerous examples of computable numbers including π, e, all the real algebraic numbers and also some general classes such as:[10]

> The sum of a power series whose coefficients form a computable sequence is a computable function in the interior of its interval of convergence.

An excellent description of the type of argument given under part (a) can be found on pages 376–381 of Kleene's classic book, *Introduction to Metamathematics* [91], where Kleene also analyzes the restrictions of a human computer. Kleene then considers four areas in which a human computer is less restricted than a machine:[11]

a) He can observe more than one symbol occurrence at a time.

b) He can perform more complicated atomic acts than the machine.

---

[10]See [160, p. 256].

[11]See [91, p. 377].

    c) His symbol space need not be a one-dimensional type.

    d) He can choose some other symbolic representation of the arguments
and function values than that used in our definition of computability.

He then shows that each area can be reduced to an equivalent in terms of
a Turing machine.

Although Turing did not explicitly state the converse of his thesis it is
obvious that he considered it to be included as part of his exposition. A proof
of the converse of Turing's thesis can be found in [**34**, p. 11].

7.5. The Non-Existence of Certain Machines

Turing adds significantly to the importance of his paper by considering some
unsolvability results. These unsolvability results include two general problems
relating specifically to machines and then, using these results, he proves that
the decision problem of the Hilbert and Ackermann functional calculus is unsolvable.

We have previously seen that to each machine there corresponds at least
one description number and that each description number corresponds to a
unique machine. Since each machine can produce at most one sequence then
the computable sequences must be enumerable.

In Section 8 of his paper Turing considers a diagonal argument that seems
to prove that the computable sequences are not enumerable. He starts by
supposing that α_n is the nth computable sequence in the enumeration, if such
an enumeration exists, and goes on to define $\phi_n(m)$ as the mth figure in α_n. He
then investigates the sequence β which has $1 - \phi_n(n)$ as its nth term.

The argument that seems to prove, by contradiction, that the computable sequences are not enumerable relies on β being assumed computable but this assumption requires that there should be a finite procedure for enumerating the computable sequences. The problem of enumerating computable sequences is recognized by Turing to be equivalent to the problem of determining whether or not a given number is the description number of a circle-free machine and, as Turing is about to prove, this can not be done in a finite number of steps.

What Turing proves is that there is no machine which, when supplied with the description number of a machine, is able to test this number and determine whether it represents a circle-free machine or a circular machine.[12] Turing achieves this by using a proof by contradiction. He assumes that such a machine $\mathscr{D}$ does exist and combines it with the universal machine to compute the sequence β' whose n^{th} figure is $\phi_n(n)$ He demonstrates that the machine H that computes β' must be circle-free. But when the machine $\mathscr{D}$ comes to consider the number K, which is the description number of $\mathscr{H}$, then it is found that $\mathscr{H}$ must be circular. This contradiction proves that the machine $\mathscr{D}$ can not exist.

Hence, assuming Turing's thesis equating computability by machine with effective computability, Turing has shown that there is no general process for determining whether or not a number is satisfactory. Turing continues by proving that there can be no machine $\mathscr{E}$ which, when supplied with the description number of an arbitrary machine $\mathscr{M}$, will determine whether $\mathscr{M}$ ever prints a given symbol, say 0.

Again Turing proves this by contradiction. He makes the assumption that E exists and uses this to produce another machine $\mathscr{G}$ that will determine if M prints 0 infinitely often. He couples this with a similarly constructed machine that will determine if $\mathscr{M}$ prints 1 infinitely often and hence produces a machine that is capable of determining whether or not $\mathscr{M}$ prints an infinity of 0's and

---

[12]Turing uses the phrase standard description when he expounds this but by his usage he must have meant description number. See [**130**, p. 8] on this point.

1's and which can therefore determine whether or not $\mathcal{M}$ is circle-free. This contradicts the previous theorem and so $\mathcal{E}$ can not exist.

Once more assuming Turing's thesis, this means there is no general process for determining whether an arbitrary machine ever prints some given symbol.

Turing uses the first of these theorems to produce a non-computable sequence δ. δ is defined as a sequence whose n^{th} figure is 1 or 0 according to whether n is or is not satisfactory. By the previous result δ is not computable. It is possible, of course, that a certain number of the digits of δ might be calculable but this would not be by any uniform, process.

Finally Turing gives an outline of his proof that there is no general process for determining whether an arbitrary formula of the Hilbert and Ackermann functional calculus is provable. That is, Turing shows that the Hilbert Entscheidungsproblem has no solution.[13]

Turing accomplishes this by constructing a formula $U_n(\mathcal{M})$ of the Hilbert and Ackermann functional calculus for each machine $\mathcal{M}$. This formula can be interpreted to mean:

> ... in some complete configuration of $\mathcal{M}$, S_1 (i.e. 0) appears on the tape.[14]

He then proves that:

a) If S_1 appears on the tape in some complete configuration of $\mathcal{M}$, then $U_n(\mathcal{M})$ is provable.

---

[13]Note that this refers to the provability form of the Entscheidungsproblem, that is, the first form designated by Church in [**22**]

[14]See [**160**, p. 260].

b) If $U_n(\mathscr{M})$ is provable, then S_1 appears on the tape in some complete configuration of $\mathscr{M}$.

Turing is then able to prove his final theorem. He assures that a general method exists for determining whether or not a given formula of the Hilbert and Ackermann functional calculus is provable. This means that there is a general procedure for determining whether or not the formula $U_n(\mathscr{M})$ is provable. If the procedure shows that $U_n(\mathscr{M})$ is provable then, by result (b), $\mathscr{M}$ must print S_1; if the procedure shows that $U_n(\mathscr{M})$ is not provable then $\mathscr{M}$ could not print S_1 or else result (a) would make $U_n(\mathscr{M})$ provable. Results (a) and (b) ensure that a general procedure for determining the provability of $U_n(\mathscr{M})$ implies a general procedure for determining whether or not $\mathscr{M}$ prints S_1.

We have already seen that there can be no machine that can test $\mathscr{M}$ and determine whether or not it will print the symbol S_1 and hence, by Turing's thesis, there can be no general procedure capable of doing this. We must therefore conclude that there is no general method capable of determining of a formula of the Hilbert and Ackermann functional calculus whether or not it is provable. Hence the Entscheidungsproblem is not solvable.

7.6. Computable, Recursive and λ-Definable Functions

Turing received an abstract of Church's 1936 paper just before he sent his own paper for publication. He immediately realized that Church's formal equivalent to effective calculability, λ-definable functions, must be equivalent to his own computability and set about to prove it. He gives an outline of the proof in the appendix to his 1936 paper [160], and goes on to give a detailed proof in his 1937 paper [161].

In the appendix of the paper, in keeping with the rest of the 1936 paper, Turing does not deal with functions but uses sequences instead. Hence Turing first defines what is meant by a sequence γ by being λ-definable:[15]

> We shall say that a sequence γ whose n^{th} figure is $\phi_\gamma(n)$ is λ-definable or effectively calculable if $1 + \phi_\gamma(u)$ is a λ-definable function of n.

Turing uses $1 + \phi_\gamma(u)$ rather than $\phi_\gamma(u)$ because he was using the Church-Kleene theory of λ-definable functions as formulated in Kleene's 1935 paper. In this paper, as in the original paper of Church, λ-definable functions were defined for positive integers only so the number 0 was not available and therefore $\phi_\gamma(u)$, whose value is 0 or 1, can not be used on its own.

In his 1937 paper, however, Turing uses $\lambda - K$-definability which includes 0 amongst its definitions, 0 being represented by $\lambda fx.x$. Also in this paper Turing uses functions rather than sequences and consequently he needs to define computable functions:[16]

> $[f(n)$ is computable if] the sequence γ_f in which there are $f(n)$ figures 1 between the nth and the $(n + l)^{st}$ 0, and $f(0)$ figures before the first 0 is computable.

The only difference between this definition and the one he indicates in his 1936 paper is that the function is defined for argument 0 as well as for all the positive integers.

In the appendix to his 1936 paper Turing proceeds to compare computable sequences of 0's and 1's with λ-definable sequences of 0's and 1's. In the proof that all λ-definable sequences are computable Turing indicates that the best

---

[15]See [**160**, p. 263].
[16]See [**161**, p. 160].

method is to define a choice machine first and then construct an automatic machine from this choice machine. He had already used this procedure earlier in the paper while considering provable formulæ of the Hilbert and Ackermann, functional calculus. At that point in the paper he remarked that once the choice machine had been constructed to produce the provable formulæ then the automatic machine capable of enumerating all the provable formulæ could be constructed from it by the following method:[17]

> [In the choice machine] we can suppose that the choices are always choices between two possibilities 0 and 1. Each proof will then be determined by a sequence of choices $i_1, \ldots, i_n$ where ($i_1=0$ or 1, $i_2=0$ or 1, $\ldots i_n=0$ or 1), and hence the number
>
> $$2^n + i_1 2^{n-1} + i_2 2^{n-2} + \ldots + i_n$$
>
> completely determines the proof. The automatic machine carries out successively proof 1, proof 2, proof 3 $\ldots$

Further on in the appendix Turing argues that the same technique can be used to produce an automatic machine that can construct successively all the formulæ into which a given formula M is convertible.

In his 1937 paper Turing goes into the details of how the choice machine would work and hence how an automatic machine can be constructed to enable the conversion process to be performed on a Turing machine. The comments on choice machines in his 1936 paper and the details in his 1937 paper represent the first actual use of choice machines.

The rest of the proof that all λ-definable sequences are computable follows fairly easily from this work.

---

[17]See [**160**, footnote p. 252].

In the 1936 paper Turing continues by showing that all computable sequences are λ-definable. A key result needed for this proof is Turing's assertion that the function ρ_γ is λ-definable (although he admits that he will omit the proof), where ρ_γ appears in the relationship:

$$\xi(n+1) = \rho_\gamma(\xi(n))$$

where $\xi(n)$ is some description number of the nth complete configuration of $\mathcal{M}$, the machine that constructs the sequence γ. That is, ρ_γ is the functional relationship between successive complete configurations of $\mathcal{M}$.

In his 1937 paper Turing proves, instead, that computable functions are recursive and then utilizes the existing proof of equivalence between recursive functions and λ-definable functions to complete his required proof of the equivalence between computable functions and λ-definable functions. In this paper Turing does go into the details of why the function that describes the relationship between successive complete configurations of $\mathcal{M}$ is recursive. It is then a fairly straightforward task to complete the required proof of the equivalence.

The proof of the equivalence theorems in Turing's 1936 and 1937 papers made it evident that Church's thesis and Turing's thesis were equivalent.

7.7. Conclusion

We have traced the gradual unfolding and rise in importance of the class of general recursive functions. As Post states, somewhat extravagantly, in his 1944 paper:[18]

> Indeed, if general recursive function is the formal equivalent
> of effective calculability, its formulation may play a role in the

---

[18]See [**129**, p. 315].

history of combinatory mathematics second only to that of the formulation of the concept of natural number.

Originally recursive definitions were used by Dedekind and Peano as convenient methods of defining the basic operations of elementary number theory.

Then Skolem, in 1923, made use of the constructive nature of recursive functions by including them as one of the basic notions in his foundation of a finitist system of arithmetic. He made extensive use of the ordinary recursive definition and extended the type of recursion used to include course-of-values recursion.

About the same time the finitary nature of recursive functions caused Hilbert to use them as the cornerstone of his metamathematical approach to the proof of the consistency of arithmetic. At this stage recursive functions were still being used piecemeal, the class of recursive functions had not been investigated in sufficient detail for it to be possible to determine precisely which definitions produced the same class of functions although Ackermann's function was known to lead out of the class of ordinary recursive functions.

This situation was resolved in the 1930's when Gödel, in 1931, gave a precise definition of the class of ordinary recursive functions and Péter, in a series of papers in the early 1930's, analyzed and organized the whole topic of the different types of recursive definition.

Gödel referred to the constructive nature of recursive functions and, after defining a decidable relation, he noted that all recursive relations were decidable. In his lectures at Princeton Gödel augmented this by making reference to the fact that primitive recursive functions were constructive in that they could be computed by a finite procedure and also that primitive recursive relations were finitely decidable since their representing functions were computable.

Meanwhile, in Princeton, Church and Kleene were pursuing a line of enquiry that was ultimately to lead to the statement of Church's thesis which linked together the different strands of development in foundational research that were taking place in the early 1930's. Church originally started his research in Princeton by attempting to produce a consistent system of arithmetic but when Kleene and Rosser proposed to prove it inconsistent he began to stress the importance of the class of λ-definable functions that could be extracted from his system.

The wealth of information pertaining to these functions that had already been produced, mainly by Kleene, led Church to suspect that they might encompass all the effectively calculable functions. He spoke to Gödel about his proposal to identify the classes of effectively calculable functions and λ-definable functions and, as we have seen, Gödel was far from happy with this identification.

Subsequently this discussion prompted Gödel to modify Herbrand's attempt at generalizing the definition of a recursive function and so to produce the class of general recursive functions. He then proceeded to introduce these functions into his lectures at Princeton.

Kleene developed the theory of these general recursive functions on the formal side and produced his very important normal form theorem that made it immediately evident to what extent they represented a generalization over the primitive recursive functions. Church's thesis equated both the general recursive functions and the λ-definable functions with the effectively calculable functions.

Kleene's results, linking decision problems with the existence of a normal form for a λ-definable formula and the final statement of Church's thesis led to the proofs of the first unsolvable decision problems in logic. Hartley Rogers, Jr., summarized these rapid developments as follows:[19]

> [the] demonstration of the existence of easily described recursively unsolvable problems is one of the more striking achievements of twentieth-century mathematics.

Even at the time of the creation of the λ-definable functions and the general recursive functions it was recognized that the general recursive functions were more in the line of familiar mathematical development.

Consequently they were preferred in exposition as can be seen from this quotation from Kleene:[20]

> I myself, perhaps unduly influenced by rather chilly receptions from audiences around 1933–35 to disquisitions on λ-definability, chose, after general recursiveness had appeared, to put my work in that format ... I thought general recursiveness came the closest to traditional mathematics. It spoke in a language familiar to mathematicians, extending the theory of special recursiveness, which derived from formulations of Dedekind and Peano in the mainstream of mathematics.
>
> I cannot complain about my audiences after 1935, although whether the improvement came from switching I do not know. In retrospect, I now feel it was too bad I did not keep active in λ-definability as well.

The final development that we have considered is another equivalent class of functions, the class of Turing computable functions. The major importance of Turing's work was the weight that his concept lent to the probable truth of

---

[19]See [**135**, p. 26].
[20]See [**94**, p. 62].

both his own thesis and that of Church. Many mathematicians found Turing's notion of computable functions a much more satisfactory and obvious equivalent to effectively calculable functions. This opinion can be found in Turing's own 1937 paper:[21]

> The identification of 'effectively calculable' functions with computable functions is possibly more convincing than an identification with the λ-definable or general recursive functions. For those who take this view the formal proof of equivalence [between computable functions and λ-definable functions] provides a justification for Church's calculus, and allows the 'machines' which generate computable functions to be replaced by the more convenient λ-definitions.

Thus Turing felt that the equivalence that he was about to prove, between computable functions and λ-definable functions, would serve a valuable purpose. It should be remembered that when Turing wrote this paper he was in Princeton and his own belief in the convincing nature of his own formal definition of effectively calculable functions was most probably amply reinforced by Church and the others at Princeton. Church was certainly of this opinion as can be seen from the following extract from his review of Turing's paper in the 1937 volume of the *Journal of Symbolic Logic*:[22]

> Of these [notions], the first [computability by Turing machine] has the advantage of making the identification with effectiveness in the ordinary (not explicitly defined) sense evident immediately i.e. without the necessity of proving preliminary theorems.

---

[21]See [**161**, p. 153].
[22]See [**24**, p. 43].

Finally, it appears that Turing's work contributed significantly in overcoming Gödel's doubts on Church's thesis. Gödel thought that Church's identification of λ-definable functions and effectively calculable functions was unsatisfactory. Church proposed to undertake to prove that any other definition that Gödel himself could suggest would be included in λ-definability. Gödel subsequently introduced the concept of general recursive functions in his 1934 lectures at Princeton but, as he stated in a letter to Davis on 15 February 1965, he was at the time of his 1934 lectures not at all convinced that his concept of recursion comprised all possible recursions.

Gödel was certainly very careful in his judgements but, as can be seen from the following extract from the *Postscriptum* that he prepared for Davis' 1965 reprint of his own 1934 lectures, he was finally convinced by Turing's work:[23]

> In consequence of later advances, in particular of the fact that, due to A. M. Turing's work, a precise and unquestionably adequate definition of the general concept of a formal system can now be given, the existence of undecidable arithmetical propositions and the non-demonstrability of the consistency of a system in the same system can now be proved rigorously for every consistent formal system containing a certain amount of finitary number theory. Turing's work gives an analysis of the concept of mechanical procedure (alias "algorithm" or "computation procedure" or "finite combinatorial procedure"). This concept is shown to be equivalent with that of a "Turing machine".

This note was dated 3 June 1964. A similar note was added to van Heijenoort's 1967 reprint of Gödel's 1931 paper, the note can be found on page 616 of van Heijenoort's book and is dated 28 August 1963.

---

[23]See [**35**, p. 71–72].

Thus any lingering doubts that Gödel had in the generality of his 1931 results and in the likely truth of Church's thesis were finally removed by Turing's work.

The theory of recursive functions and the notion of computability, whose beginnings we have traced thereafter became widely recognized and accepted. Kleene, Church, Turing and many others continued the work of development and all of these classes of functions have continued to figure in many important mathematical papers to the present day.

Dates of Major Figures

Ackermann, Wilhelm	29 March 1896 – 24 December 1962
Bernays, Paul	17 October 1888 – 18 September 1977
Church, Alonzo	14 June 1903 – 8 November 1995
Dedekind, Richard	6 October 1831 – 12 February 1916
Frege, Friedrich Ludwig Gottlob	8 November 1848 – 26 July 1925
Gödel, Kurt	28 April 1906 – 14 January 1978
Grassmann, Hermann Günther	15 April 1809 – 26 September 1877
Herbrand, Jacques	12 February 1908 – 27 July 1931
Hilbert, David	23 January 1862– 14 February 1943
Kalmár, László	27 March 1905 – 2 Aug 1976
Kleene, Stephen Cole	5 January 1909 – 25 January 1994
von Neumann, John	28 December 1903 – 8 February 1957
Peano, Giuseppe	27 August 1858 – 20 April 1932
Peirce, Charles Sanders	10 September 1839 – 19 April 1914
Péter, Rózsa	17 February 1905 - 16 February 1977
Post, Emil Leon	11 February 1897 – 21 April 1954
Skolem, Thoralf	23 May 1887 – 23 March 1963
Turing, Alan Mathison	23 June 1912 – 7 June 1954

APPENDIX B

Letters

This appendix contains eleven letters of historical interest. The first letter is from Alonzo Church to Stephen C. Kleene and was enclosed in the letter of November 3, 1981, from Kleene to the author. The other ten are to the author.

The table below lists the letters including the author of the letter, the date of the letter and a brief description of some attributes the letter. Each of the following sections contains a scanned image of a letter followed by a verbatim transcription.

Author	Date	Description
Author	*Date*	*Description*
Alonzo Church	November 29, 1935	typewritten, two pages, Princeton letterhead
Stephen C. Kleene	July 28, 1977	typewritten, three pages, handwritten postscript, U W Madison letterhead
Alonzo Church	April 21, 1978	handwritten, one page, posted from Los Angeles
Hao Wang	July 16, 1979	handwritten, one pages, The Rockefeller University letterhead
Stephen C. Kleene	July 17, 1979	typewritten, two pages, UW Madison
Jean van Heijenoort	August 17, 1979	handwritten, two pages, posted from Cambridge, MA
Stephen C. Kleene	November 3, 1981	typewritten, seven pages plus one page chronology, U W Madison letterhead
Stephen C. Kleene	November 13, 1981	typewritten, two pages, UW Madison letterhead
J. Barkley Rosser	November 19, 1981	handwritten, one page, personal UW letterhead
Stephen C. Kleene	September 9, 1982	four pages, typewritten, UW Madison letterhead
Stephen C. Kleene	April 4, 1983	two pages, typewritten, posted from Madison, WI

B.1. Alonzo Church to Stephen C. Kleene, November 29, 1935

PRINCETON UNIVERSITY

PRINCETON NEW JERSEY

Department of
MATHEMATICS

November 29,1935.

Dear Kleene:

I have your letter of November eighteenth.

The notion of lambda-definability in its present form is, of course, the result of a gradual developement. The first step was my proposal, within the system of formal logic in which I was working, of definitions of the positive integers and of the function S, and remark as to how, in terms of these, a definition by recursion could (at least in certain cases) be translated into a nominal definition, capable of being formalized in the system. The next steps, taken by you, were the restriction to consideration of a part only of my formal system (so that conversion became the only available way of proving equality), the developement in the direction of finding general proofs of definability for various classes of functions, the discovery of such powerful instruments of definition as the perpetual motion function, and the remark that the question of the truth of certain propositions of elementary number theory is reducible to the question whether certain particular formulas have a normal form. After these, however, two steps which I think to be of some importance were taken by myself. The first of these was the proposal of the problem of a complete set of invariants of conversion and the remark that the solution of it would imply a solution of the Entscheidungsproblem of Principia Mathematica; this was made not long after our original discussion of the perpetual motion function and was, of course, immediately suggested by that discussion. The second was the proposal to abstract from any formal system of logic and to consider the class of lambda-definable functions of positive integers as a problem of number theory; this was first made to a number of persons in the fall of 1933 in connection with Rosser's discovery of the contradiction in my system, and afterwards to you in connection with the consequent revision of your thesis.

At any rate, we seem to be agreed that the statement that the notion of lambda-definability is jointly due to you and me is fair, and I am content to let it go at that.

In regard to Gödel and the notions of recursiveness and effective calculability, the history is the following. In discussion with him the notion of lambda-definability, it developed that there was no good definition of effective calculability. My proposal that lambda-definability be taken as a definition of it he regarded as thoroughly unsatisfactory. I replied that if he would propose any definition of effective calculability which seemed even partially satisfactory I would undertake to prove that it was included in lambda-definability. His only idea at the time was that it might be possible, in terms of effective calculability as an undefined notion, to state a set of axioms which would embody the generally accepted properties of this notion, and to do something on that basis. Evidently it occurred to him later that Herbrand's definition of recursiveness, which had no regard to effective calculability, could be modified in the direction of effective calculability, and he made this proposal in his lectures. At that time he did specifically raise the question of the connection between recursiveness

Dr. S. C. Kleene   11/29/35

in this new sense and effective calculability, but said that he did not
think that the two ideas could be satisfactorily identified "except heuristically".

In regard to your functions H, $Eval_p$, and Val, it seems to me
that it might be possible to obtain for them definitions which were not
really recursive at all but actual nominal definitions in terms of addition, multiplication, and the epsilon operator. Such definitions would
be, in a sense, less simple than primitive recursions, since they would
involve the epsilon operator. But they would lead to the interesting
theorem (if it be a theorem) that for any general recursive function it
is possible to give a nominal definition in terms of addition, multiplication, and the epsilon operator.

Gödel is going back to Europe this week. He has been suffering
severely with indigestion ever since his arrival, and concluded finally to
resign his appointment with the Institute and go home.

I do not expect to be at the St. Louis meeting. Possibly Bernays
will attend it, I do not know.

                              Sincerely yours,

                              Alonzo Church

                              Alonzo Church.

Dr. S. C. Kleene,
University of Wisconsin,
Madison, Wisconsin.

Dear Kleene:

I have your letter of November eighteenth.

The notion of lambda-definability in its present form is, of course, the result of a gradual development. The first step was my proposal, within the system of formal logic in which was working, of definitions of the positive integers and of the function S, and remark as to how, in terms of these, a definition by recursion could (at least in certain cases) be translated into a nominal definition, capable of being formalized in the system. The next steps, taken by you, were the restriction to consideration of a part only of my formal system (so that conversion became the only available way of proving equality), the development in the direction of finding general proofs of definability for various classes of functions, the discovery of such powerful instruments of definition as the perpetual motion function, and the remark that the question of the truth of certain propositions of elementary number theory is reducible to the question whether certain particular formulas have a normal form. After these, however, two steps which I think to be of some importance were taken by myself. The first of these was the proposal of the problem of a complete set of invariants of conversion and the remark that the solution of it would imply a solution to the Entscheidungsproblem of *Principia Mathematica;* this was made not long after our original discussion of the perpetual motion function and was, of course, immediately suggested by that discussion. The second was the proposal to abstract from any formal system of logic and to consider the class of lambda-definable functions of positive integers as a problem of number theory; this was first made to a number of persons in the fall of 1933 in connection with Rosser's a discovery of the contradiction in my system, and afterwards to you in connection with the consequent revision of your thesis.

At any rate, we seem to be agreed that the statement that the notion of lambda-definability is jointly due to you and me is fair, and I am content to let it go at that.

In regard to Gödel and the notions of recursiveness and effective calculability, the history is the following. In discussion with him the notion of lambda-definability, it developed that there was no good definition of effective calculability. My proposal that lambda-definability be taken as a definition of it he regarded as thoroughly unsatisfactory. I replied that if he would propose any definition of effective calculability which seemed even partially satisfactory I would undertake to prove that it was included in lambda-definability. His only idea at the time was that it might be possible, in terms of effective calculability as an undefined notion, to state a set of axioms which would embody the generally accepted properties of this notion, and to do something on that basis. Evidently it occurred to him later that Herbrand's definition of recursiveness, which had no regard to effective calculability, could be modified in the direction of effective calculability, and he made this proposal in his lectures. At that time he did specifically raise the question of the connection between recursiveness in this new sense and effective calculability, but said that he did not think that the two ideas could be satisfactorily identified "except heuristically."

In regard to your functions H, $Eval_p$, and Val, it seems to me that it might be possible to obtain for them definitions which were not really recursive at all but actual nominal definitions in terms of addition, multiplication, and the epsilon operator. Such definitions would be, in a sense, less simple than primitive recursions, since they would involve the epsilon operator. But they would lead to the interesting theorem (if it be a theorem) that for any general recursive function it is possible to give a nominal definition in terms of addition, multiplication, and the epsilon operator.

Gödel is going back to Europe this week. He has bean suffering severely with indigestion ever since his arrival, and concluded finally to resign his appointment with the Institute and go home.

I do not expect to be at the St. Louis Meeting. Possibly Bernays will attend it, I do not know.

Sincerely yours,

Alonzo Church

Dr. S. C. Kleene,
University of Wisconsin,
Madison, Wisconsin.

B.2. Stephen C. Kleene to Author, July 28, 1977

Mathematics Department
University of Wisconsin—Madison

Van Vleck Hall
480 Lincoln Drive
Madison, Wisconsin 53706
Telephone: (608) 263-3053

uw madison

July 28, 1977

Mr. R. G. Adams
St. Albans College
29 Hatfield Road
St. Albans, Herts.
ENGLAND

Dear Mr. Adams:

1. I do not think I stated at Monash that von Neumann came over to Princeton in the fall of 1931. I did state that von Neumann spoke at the mathematics colloquium at Princeton in the fall of 1931, and that this was when Church and I first heard of Gödel's incompleteness results. *Recently* ~~today~~ I visited our library and ascertained that von Neumann was Visiting Professor of Mathematical Physics at Princeton in 1930, Professor 1931-33, Professor at the Institute for Advanced Study 1933--.† I have verified that the conference which von Neumann, Carnap, Gödel and others attended was at Konigsberg, 5-7 September, 1930.

2. Church had already submitted his first paper on "A set of postulates for the foundation of logic" before this colloquium of von Neumann's. I recollect a manuscript of Church's second paper thereon as being available to the class, or at least parts of its contents as being presented to the class in 1931-32I, before or at least independently of, the von Neumann colloquium.

3. Church had lectured on his efforts, which seemed to come tantalizingly close to success, to prove the consistency of his postulates (just one case of an induction failed to go through). I recall Church expressing the view that his foundation of logic was sufficiently different from that in PM and related systems for which Gödel had proved his (second) theorem, that he thought it possible (or likely?) that his foundation would escape from the (second) Gödel theorem. I can only say that Church expressed this view very shortly after we learned of Gödel's results through von Neumann's talk. (Of course, we then speedily set to work to read the 1931 Gödel paper itself.) Whether he held this view for any length of time thereafter I do not know. (More on this at the end of the letter.)

4. The specific event that caused me to revise my thesis, "A theory of positive integers in formal logic," which ~~was~~ *then some time later* actually in press,* was the discovery by Rosser and me that Church's system is inconsistent. We suspected the inconsistency already in January or February 19~~77~~ perhaps by the end of March or April we had pretty well confirmed the fact. (1934†)

Mr. R.G. Adams
Page 2
July 28, 1977

The thesis problem Church had given me around the beginning of February 1932 was
to develop the theory of positive integers in his system, starting from the
beginnings given in his second paper§9 (which was then available to me, or at
least this part of it).

It was only in the progress of doing so that the idea, and success, of defining
functions of positive integers entirely in the λ-calculus (without the descriptive
operator ι) arose. You may notice that, in§9, besides the fundamental successor
function $\underline{S}$, Church gives only one further example of a λ-definable function (the
term "λ-definable" I believe was first used by Church in his 1936 paper), namely ⚹ +.
He defines – using the operator ι, and thence **X**. In February 1934, I think it was,
I showed that $\underline{x}$-1 or $\underline{P}(\underline{x})$ (the predecessor function) is λ-definable; when I did so,
he said he had previously just about convinced himself that there is no λ-definition
of the predecessor function. So you see the scope of λ-definability was not
planned or anticipated, but revealed itself in the investigation rather to his
surprise. Only after this revelation was the question asked by Church whether all
effectively calculable functions are λ-definable.

When Rosser and I had convinced ourselves that Church's system is inconsistent,
the interest is developing the theory of positive integers in it (Church, second
paper, p. 804) was punctured. I then recalled my 1934 paper and my thesis
(1935), including part already in printers proof, and recast them to retain (1)
those parts only of the development of the theory of positive integers which would
be needed in the proof of inconsistency, (2) the theory of λ-definability,
considered separately from Church's foundation of logic, which had become of interest
in its own right.

I do not think these events were in any way the result of my meeting Gödel and
learning of his notion of general recursive functions. Rosser and I were already
engaged in proving Church's system inconsistent just before I met Gödel; and the
interest in λ-definability, and Church's speculation whether it might encompass
all effectively calculable functions, already existed when we heard from Gödel
of his (or the Herbrand-Gödel) general recursive functions.

5. In my paper "General recursive functions of natural numbers," I obtained some
results on classes of numbers which can be enumerated by recursive functions (or
even by primitive recursive functions), and also theorems (XI, XVI) about classes
which are not "recursively enumerable." These results were original; such
questions had not, to my knowledge, been asked and answered before. (And there
really was no "before" for general recursive functions, except Gödel's 1934 lectures
introducing the concept.) So I suppose I invented the concept of "recursively
enumerable sets." You must remember that Emil Post in 1944, p. 308, tidied the
concept up by including the empty set as also "recursively enumerable."

6. I set out, during the academic year 1934-35, after completing my collaboration
with Rosser and the final editing of my 1934 and 1935 papers ("Proof by cases ⟋ ⸌⸍ ",
"A theory of positive integers..."), to prove the equivalent of the λ-definable
and the general recursive functions. I shortly realized that the easiest way
to get the λ-definability of all general recursive functions was first to get my
normal from ψ(εyR[φ,y]) for general recursive functions. So my paper "General

Mr. R.G. Adams
Page 2
July 28, 1977

recursive functions..." was a spin-off from "λ-definability and recursiveness."
I recognized at once, I believe, that the normal form, and the other results I put
into that paper, would make it a more interesting paper than "λ-definability...", to
support which I had started on it. It was natural for me to involve the least-number
operator (then written "εy"), since this had played a central role in my work on
λ-definability, where (p. 231) I called it the "β"-function (German l.c. "β" for
"perpetual notion").   As it was really the natural descriptive operator for working
with the positive integers or the natural numbers, I could hardly have progressed
far in λ-definability without it. Hilbert-Bernays, Vol. I (1934), likewise found
their least-number operator μ$_x$ ~~indispensable~~ (p. 396). ✱✱
                                very useful!

An important subsequent step, taken in my 1938 paper "On notation...", was to go
over to partial functions, using a least-number operator μy taken to be undefined
when the sought y does not exist. (The "β" in λ-definability was used this way.)

2 and 3 (additional). Your question suggests you may be wondering whether Gödel's
1931 results influenced Church's two papers, "A set of postulates for the foundation
of logic." I am sure Church's work in those two papers was not in any significant
way influenced by Gödel 1931. I now find on looking in the second paper, the
top of p. 843, that Church actually suggested in print that his system might escape
Gödel's second theorem; so he did persist in the view I have reported in 3 for
some time (at least till January 3, 1933). The first paper grew out of work Church
did as a National Research Fellow in 1928-29; and the momentum of this research
could hardly but carry him along the course followed in the second paper.

                              Sincerely yours,

                              Stephen C. Kleene

                              Stephen C. Kleene

SCK/deb

P.S. ✱N.B. On reflection, I do not remember whether Part I of this, or type-set
      my 1934 paper, or both, were then ~~in print~~ type set, some type-set
      material had to be reset.
      † The last previous entry in the old volume of "Who's
Who in America" from which I got this information on
von Neumann was "Privatdozent Univ. of Berlin 1927".
      ‡ This was before we were introduced to general
recursive functions.
          ° The ~~equivalent~~ question, what classes of numbers are the range
of a λ-definable function, of course could have been considered
before then, but was not.
          ✱✱ Consider the setting in which I discovered my normal
form for general recursive functions: I had already known for
nearly two years before I heard of general recursive functions
that ~~class of the~~ the λ-definable functions is closed under primitive recursion,
                              composition, and the least-number operator.

Dear Mr. Adams:

1. I do not think I stated at Monash that von Neumann came over to Princeton in the fall of 1931. I did state that von Neumann spoke at the mathematics colloquium at Princeton in the fall of 1931, and that this was when Church and I first heard of Gödel's incompleteness results. Recently I visited our library and ascertained that von Neumann was Visiting Professor of Mathematical Physics at Princeton in 1930, Professor 1931–33, Professor at the Institute for Advanced Study 1933–.[1] I have verified that the conference which von Neumann, Carnap, Gödel and others attended was at Königsberg, 5–7 September, 1930.

2. Church had already submitted his first paper on "A set of postulates for the foundation of logic" before this colloquium of von Neumann's. I recollect a manuscript of Church's second paper thereon as being available to the class, or at least parts of its contents as being presented to the class in 1931-32I [sic], before or at least independently of, the von Neumann colloquium.

3. Church had lectured on his efforts, which seemed to come tantalizingly close to success, to prove the consistency of his postulates (just one case of an induction failed to go through). I recall Church expressing the view that his foundation of logic was sufficiently different from that in PM and related systems for which Gödel had proved his (second) theorem, that he thought it possible (or likely?) that his foundation would escape from the (second) Gödel theorem. I can only say that Church expressed this view very shortly after we learned of Gödel's results through von Neumann's talk. (Of course, we then speedily set to work to read the 1931 Gödel paper itself.) Whether he held this view for any length of time thereafter I do not know. (More on this at the end of the letter.)

---

[1]The last previous entry in the old volume of "Who's Who in America" from which I got this information on von Neumann was "Privatdozent U. of Berlin 1927.

4. The specific event that caused me to revise my thesis, "A theory of positive integers in formal logic," which may have been then actually in press [2], was the discovery by Rosser and me that Church's system is inconsistent. We suspected the inconsistency already in January or February 1934[3], perhaps by the end of March or April we had pretty well confirmed the fact.

The thesis problem Church had given me around the beginning of February 1932 was to develop the theory of positive integers in his system, starting from the beginnings given in his second paper §9 (which was then available to me, or at least this part of it).

It was only in the progress of doing so that the idea, and success, of defining functions of positive integers entirely in the λ-calculus (without the descriptive operator $\imath$) arose. You may notice that, in §9, besides the fundamental successor function $\underline{S}$, Church gives only one further example of λ-definable function (the term "λ-definable" I believe was first used by Church in his 1936 paper), namely + [plus]. He defines − [minus] using the operator $\imath$, and thence × [multiplication]. In February 1934, I think it was, I showed that $\underline{x} \dot{-} 1$ or $\underline{P}(\underline{x})$ (the predecessor function) is λ-definable; when I did so, he said he had previously just about convinced himself that there is no λ-definition of the predecessor function. So you see the scope of λ-definability was not planned or anticipated, but revealed itself in the investigation rather to his surprise. Only after this revelation was the question asked by Church whether all effectively calculable functions are λ-definable.

When Rosser and I had convinced ourselves that Church's system is inconsistent, the interest is developing the theory of positive integers in it (Church, second paper, p. 804) was punctured. I then recalled my 1934 paper and my thesis (1935), including part already in printers proof, and recast them to retain

---

[2]N.B. On reflection I do not remember whether Part I of this, or my 1934 paper, or both, were then typeset. Some type set material had to be reset.

[3]This was before we were introduced to general recursive functions

(1) those parts only of the development of the theory of positive integers which would be needed in the proof of inconsistency, (2) the theory of λ-definability, considered separately from Church's foundation of logic, which had become of interest in its own right.

I do not think these events were in any way the result of my meeting Gödel and learning of his notion of general recursive functions. Rosser and I were already engaged in proving Church's system inconsistent just before I met Gödel; and the interest in λ-definability, and Church's speculation whether it might encompass all effectively calculable functions, already existed when we heard from Gödel of his (or the Herbrand-Gödel) general recursive functions.

5. In my paper "General recursive functions of natural numbers," I obtained some results on classes of numbers which can be enumerated by recursive functions (or even by primitive recursive functions), and also theorems (XI, XVI) about classes which are not "recursively enumerable." These results were original; such questions had not, to my knowledge, been asked and answered before. (And there really was no "before" for general recursive functions, except Gödel's 1934 lectures introducing the concept.[4]) So I suppose I invented the concept of "recursively enumerable sets." You must remember that Emil Post in 1944, p. 308, tidied the concept up by including the empty set as also "recursively enumerable."

6. I set out, during the academic year 1934–35, after completing my collaboration with Rosser and the final editing of my 1934 and 1935 papers ("Proof by cases ...", "A theory of positive integers ..."), to prove the equivalent of the λ-definable and the general recursive functions. I shortly realized that the easiest way to get the λ-definability of all general recursive functions was first to get my normal form $\psi(\epsilon y R[\phi, y])$ for general recursive functions. So my paper "General recursive functions ..." was a spin-off from "λ-definability and

---

[4]The equivalent question, what classes of numbers are the range of a λ-definable function, of course could have been considered before then, but was not.

recursiveness." I recognized at once, I believe, that the normal form, and the other results I put into that paper, would make it a more interesting paper than "λ-definability . . . ", to support which I had started on it. It was natural for me to involve the least-number operator (then written "εy"), since this had played a central role in my work on λ-definability, where (p. 231) I called it the "$\mathfrak{p}$"-function (German 1. c. "$\mathfrak{p}$" for "perpetual notion"). As it was really the natural descriptive operator for working with the positive integers or the natural numbers, I could hardly have progressed far in λ-definability without it. Hilbert-Bernays, Vol. I (1934), likewise found their least-number operator μ_x very useful (p. 396). [5]

An important subsequent step, taken in my 1938 paper "On notation . . . ", was to go over to partial functions, using a least-number operator μy taken to be undefined when the sought y does not exist. (The "$\mathfrak{p}$" in λ-definability was used this way.)

2 and 3 (additional). Your question suggests you may be wondering whether Gödel's 1931 results influenced Church's two papers, "A set of postulates for the foundation of logic." I am sure Church's work in those two papers was not in any significant way influenced by Gödel 1931. I now find on looking in the second paper, the top of p. 843, that Church actually

---

[5]Consider the setting in which I discovered my normal form for general recursive functions: I had already known for nearly two years before I heard of general recursive functions that the class of the λ-definable functions is closed under primitive recursion, composition, and the least-number operator.

suggested in print that his system might escape Gödel's second theorem; so he did persist in the view I have reported in 3 for some time (at least till January 3, 1933). The first paper grew out of work Church did as a National Research Fellow in 1928–29; and the momentum of this research could hardly but carry him along the course followed in the second paper.

Sincerely yours,

Stephen C. Kleene

SCK/deb

B.3. Alonzo Church to Author, April 21, 1978

Los Angeles, April 21, 1978

Dear Mr. Adams,

My apologies for a much delayed reply to your letter. I do have great difficulty in keeping up with correspondence, my replies are often delayed, and it seems that your letter must have been somehow mislaid while awaiting an answer.

Answers to many of your questions are in a paper by S.C. Kleene which appeared in volume 41 number 4 of The Journal of Symbolic Logic (December 1976), and I hope that by this time you may have seen and studied this paper.

I believe that I first heard of Gödel's incompleteness theorem indirectly in late 1931. It was certainly unknown to me at the time that my first paper in the Annals of Mathematics was submitted, and it now seems that even at the time of the second paper I continued to hope for more in the direction of partial consistency proofs and relative consistency proofs than ought to have seemed reasonable. (However, for the other side of the coin, see my paper in Proceedings of the Tarski Symposium, published in 1974.)

The notion of a general recursive function is due to Gödel, in his lectures of 1933 and 1934 to which Kleene refers. (There had indeed been a previous suggestion by Herbrand, but his formulation is far from adequate, and the notion must be credited rather to Gödel.) The firm thesis that general recursiveness is more than just a generalization of what is now called primitive recursiveness and is to be identified with the informal notion of effective calculability is mine (in an abstract in Bulletin of the American Mathematical Society in May 1935). As Kleene points out, this thesis was suggested at least in part by his own previous very remarkable and unexpected results in the theory of λ-definable functions, and by the equivalence between general recursiveness and λ-definability which was more or less immediately evident in consequence of these results.

Very sincerely yours,

Alonzo Church

Los Angeles, April 21, 1978

Dear Mr. Adams,

My apologies for a much delayed reply to your letter. I do have great difficulty in keeping up with correspondence, my replies are often delayed, and it seems that your letter must have been somehow mislaid while awaiting an answer.

Answers to many of your questions are in a paper by S. C. Kleene which appeared in volume 41 number 4 of *The Journal of Symbolic Logic* (December 1976), and I hope that by this time you may have seen and studied this paper.

I believe that I first heard of Gödel's incompleteness theorem indirectly in late 1931. It was certainly unknown to me at the time that my first paper in the *Annals of Mathematics* was submitted, and it now seems that even at the time of the second paper I continued to hope for more in the direction of partial consistency proofs and relative consistency proofs than ought to have seemed reasonable. (However, for the other side of the coin, see my paper in *Proceedings of the Tarski Symposium,* published in 1974.)

The notion of a general recursive function is due to Gödel, in his lectures of 1933 and 1934 to which Kleene refers. (There had indeed been a previous suggestion by Herbrand, but his formulation is far from adequate, and the notion must be credited rather to Gödel.) The firm thesis that general recursiveness is more than just a generalization of what is now called primitive recursiveness and is to be identified with the informal notion of effective calculability is mine (in an abstract in *Bulletin of the American Mathematical Society* in May 1935). As Kleene points out, this thesis was suggested at least in part by his own previous very remarkable and unexpected results in the theory of λ-definable functions, and by the equivalence between general recursiveness and λ-definability which was more or less immediately evident in consequence of these results.

Very sincerely yours,

Alonzo Church

B.4. Hao Wang to Author, July 16, 1979

THE ROCKEFELLER UNIVERSITY
1230 YORK AVENUE · NEW YORK, NEW YORK 10021

July 16, 1979

Dear Mr. Adams,

Thank you for your letter dated June 27, postmarked July 6.

(1) I believe neither Hilbert nor Gödel knew of Skolem's paper when they wrote on primitive recursive fcns.

(2) On the background of Gödel's incompleteness theorems, he told me some very interesting things, I enclose a copy of a talk of mine (see the two paras. marked)

Yours sincerely,
Hao Wang

Enclosure.

July 16, 1979

Dear Mr. Adams,

Thank you for your letter dated June 27, postmarked July 6.

(1) I believe neither Hilbert nor Gödel knew of Skolem's paper when they wrote on primitive recursiveness.

(2) On the background of Gödel incompleteness theorems, he told me some very interesting things, I enclose a copy of a talk of mine (see the two paras. marked)

Very sincerely yours,

Hao Wang

Enclosure.

AUTHOR'S NOTE: The two paragraphs referred to are quoted on pages 54–55.

B.5. Stephen C.Kleene to Author, July 17, 1979

Mathematics Department
University of Wisconsin—Madison

uw madison

Van Vleck Hall
480 Lincoln Drive
Madison, Wisconsin 53706
Telephone: (608) 263-3053

July 17, 1979

Mr. R. Adams
St. Albans College
29 Hatfield Road,
St. Albans, Herts.,
ENGLAND

Dear Mr. Adams:

Thank you for your note of June 17. I should tell you that, besides my communication to you in July 1977, and my participation in "Reminiscences of Logicians" at Monash University, I also gave answers in November 1973 to questions posed by Robert L. Constable, Department of Computer Sciences, Cornell University, and I will speak at the 20th Annual Symposium on Foundations of Computer Science at San Juan, Puerto Rico, October 29-31, 1979. These four items all constitute reminiscences and historical data on the general subject we are discussing. The answers to questions received by Robert Constable may be published by him, though his publication plans seem to slip from year to year without being brought to a conclusion. My lecture at the Symposium in Puerto Rico will be published, presumably fairly promptly after October 1979.

In response to the two questions in your present letter, I doubt that Gödel knew of Skolem's 1923 paper. It seems to me just unlikely that someone with as much momentum on his own research as Gödel would have spent very much time digging around in obscure places in the literature. Of course, someone might have called Skolem's paper to his attention, but I would suspect not until later than his 1931 paper. I repeat, I am just guessing.

I have no idea whether Gödel was trying to prove arithmetic consistent when he produced his unexpected incompleteness results in the 1931 paper. He may have been imaginative enough to have pondered the nature of the problem in a broad context without actually immersing himself in a specific effort to prove consistency and have the insight he had then to get his result by applying a modification of the Paradox of the Liar. Indeed Finsler in 1926 (see the reference in my book "Introduction to Metamathematics") used much the same argument, but in a context which made it fall short of Gödel's 1931 result. Maybe Gödel even saw the Finsler paper, which was in Zeitschrift, which he is more likely to have seen than the publication having Skolem's paper in it. Gödel announced

Mr. R. Adams
Page 2
July 17, 1979

his 1931 result on 7 September 1930 at Königsberg (see Erkenntius, Vol. 2,
<u>Diskussion zur Grundlegung der Mathematik</u>), which is so soon after he
wrote his paper on completeness, that it is hard to believe he didn't
proceed quite directly to his 1931 results--at least there was hardly
time for any very extensive investment of time in trying to prove arithmetic consistent.

                                Sincerely yours,

                                Stephen C. Kleene

SCK:jmp

Dear Mr. Adams:

Thank you for your note of June 17. I should tell you that, besides my communication to you in July 1977, and my participation in "Reminiscences of Logicians" at Monash University, I also gave answers in November 1973 to questions posed by Robert L. Constable, Department of Computer Sciences, Cornell University, and I will speak at the 20th Annual Symposium on Foundations of Computer Science at San Juan, Puerto Rico, October 29–31, 1979. These four items all constitute reminiscences and historical data on the general subject we are discussing. The answers to questions received by Robert Constable may be published by him, though his publication plans seem to slip from year to year without being brought to a conclusion. My lecture at the Symposium in Puerto Rico will be published, presumably fairly promptly after October 1979.

In response to the two questions in your present letter, I doubt that Gödel knew of Skolem's 1923 paper. It seems to me just unlikely that someone with as much momentum on his own research as Gödel would have spent very much time digging around in obscure places in the literature. Of course, someone might have called Skolem's paper to his attention, but I would suspect not until later than his 1931 paper. I repeat, I am just guessing.

I have no idea whether Gödel was trying to prove arithmetic consistent when he produced his unexpected incompleteness results in the 1931 paper. He may have been imaginative enough to have pondered the nature of the problem in a broad context without actually immersing himself in a specific effort to prove consistency and thus have had the insight to get his result by applying a modification of the Paradox of the Liar. Indeed Finsler in 1926 (see the reference in my book *Introduction to Metamathematics*) used much the same argument, but in a context which made it fall short of Gödel's 1931 result. Maybe Gödel even saw the Finsler paper, which was in *Zeitschrift*, which he is more likely to have seen than the publication having Skolem's paper in it. Gödel announced his 1931 result on 7 September 1930 at Königsberg (see *Erkenntius*,

Vol. 2, "Diskussion zur Grundlegung der Mathematik"), which is so soon after he wrote his paper on completeness, that it is hard to believe he didn't proceed quite directly to his 1931 results – at least there was hardly time for any very extensive investment of time in trying to prove arithmetic consistent.

Sincerely yours,

Stephen C. Kleene

SCK:jmp

B.6. Jean van Heijenoort to Author, August 17, 1979

Pusey 207
Harvard University
Cambridge, MA 02138
U. S. A.
17 August 1979

Dear Mr. Adams,

I had in my hands your letter of 27 June with some delay, because I now go only rarely to Brandeis. Please use the address above.

I am quite certain that Hilbert and Gödel knew Skolem's 1923 paper. The paper is mentioned in Hilbert and Bernays. When I met Gödel in Princeton, he made a point of telling me that he did not know Skolem's 1922 paper, which seems to imply that he knew the 1923 paper. As to why the paper was rarely mentioned, though, I repeat, it is mentioned in Hilbert-Bernays and perhaps some other papers by Bernays — see my introductory note to the paper in From Frege to Gödel, (p. 302-303), the reason may be that it presented a point of view that was readily accepted, but no deep technical results.

The immediate stimulus for Gödel in writing his 1931 paper was the attempts to reproduce within the system the Liar's paradox and hence to define truth.

One immediate inspiration was Fraenkel's work. But
Hilbert's consistency problem was looming in the
background all the time.

Sincerely,

J. van Heijenoort

Dear Mr. Adams,

I had in my hands your letter of 27 June with some delay, because I now go only rarely to Brandeis. Please use the address above.

I am quite certain that Hilbert and Gödel know Skolem's 1923 paper. The paper is mentioned in Hilbert and Bernays. When I met Gödel in Princeton, he made a point of telling me that he did not know Skolem's *1922* paper, which seems to imply that he knew the 1923 paper. As to why the paper was rarely mentioned (though, I repeat, it is mentioned in Hilbert-Bernays and perhaps some other papers by Bernays – see my introductory note to the paper in *From Frege to Gödel*, pp. 302–303), the reason may be that it presented a point of view that was readily accepted, but no deep technical results.

The immediate stimulus for Gödel in writing his 1931 paper was the attempt to reproduce within the system the Liar's paradox and hence to define truth. One immediate inspiration was Finsler's work. But Hilbert's consistency problem was looming in the background all the time.

Sincerely,

J. van Heijenoort

B.7. Stephen C.Kleene to Author, November 3, 1981

uw madison

Mathematics Department
University of Wisconsin—Madison

Van Vleck Hall
480 Lincoln Drive
Madison, Wisconsin 53706
Telephone: (608) 263-3053

November 3, 1981

Mr. R. G. Adams
St. Albans College
29 Hatfield Road
St. Albans, Herts.
ENGLAND

Dear Mr. Adams:

Thank you for sending me a copy of the handwritten MSS of your Chapters 1-5.

In writing my two 1981 papers (reprints of which I am sending under separate cover) and my Biographical Memoir on Kurt Gödel for the National Academy of Sciences (about to be sent to press), I have come up with some more details about the events under discussion. So I have reread my letters to you of July 28, 1977 and July 17, 1979 to see what I would now change or amplify.

July 1979, third paragraph. Yes, Gödel was "study[ing] the problem of proving the consistency of analysis [and thus of arithmetic]" when he obtained his famous 1931 results, as indicated by Wang in Jour. Symbolic Logic, vol. 46 (1981), pp. 654-655. And Gödel did know of Finsler's 1926 paper, according to Kreisel in Zentralblatt, vol. 401 (1979), p. 13.

July 1977, second ⁋ on p. 2. "February 1934" is a misprint for "February 1932", as you must have realized. The history related in the last paragraph of p. 1 is more accurately as follows.

Rosser came up with the idea of proving the inconsistency of Church's system (with an outline of the reasoning to be used) in the fall of 1933, but of course without knowing instantly whether or not his proposed reasoning could be carried out in full detail. He told his proposed plan for proving the inconsistency to Church, who reacted in two ways. First, he spoke out on the significance of λ-definability as a notion of number theory of interest in its own right, which he "afterwards" communicated "to [me] in connection with the consequent revision of [my] thesis." I am quoting from his letter to me of November 29, 1935 (copy enclosed). I was off in Maine, so I'm not sure just how early I learned about all this. I remember that in December 1933 or January 1934, Rosser wrote me his plan for proving the inconsistency of Church's system and recruited me to join him in carrying it out. Of course, carrying out the proof called for expertise in developing theory within Church's system, and this is just what I had. I remember working on it, some weeks, in my farmhouse in Maine, and then being offered a Research Assistantship to continue the work at Princeton, where I could work more efficiently. Of

-2-

course, in Church's letter "Rosser's discovery of the contradiction in [his] system" means the discovery of the idea for proving a contradiction in the system. It took some time to actually formalize Rosser's sketch, and our confidence that the proof could be done no doubt gradually increased as I ran into no snags in my research on it. We were probably pretty sure the proof would go through "by end of March or April", as I wrote. Church's other reaction to learning of Rosser's plan in the fall of 1933 was to write, deliver and publish, his paper, The Richard Paradox. As stated above, I was in Maine while this was going on. My return to Princeton was on February 7, 1934. But Rosser told me then very explicitly that writing The Richard Paradox was a response by Church to his [R's] proposal of his plan to derive a contradiction in Church's system.

I enclose a Chronology that I recently had another occasion to compose. The events described in my Annals of the History of Computing, vol. 3 (1981) paper on p. 59 left column lines 4-1 f.b. and right column lines 8-14 (in the second paragraph) belong in the period shown by the box in my Chronology.

Evidently, when I wrote you in July 1977, I had forgotten both about Church's letter of November 29, 1936, and about Church's paper The Richard Paradox (which had no effect on my research) with its December 30, 1933 date of delivery to the A.M.S., both of which put Rosser's discovery of his plan well back into the fall of 1933. It was easier for me to forget these, since I was not in Princeton when the events took place.

Now I will comment on points in your MSS.

Chapter 1, p. 17. You cite Zermelo (1908) and Fraenkel (1922). Should you also mention Skolem (1922) and Weyl (1910), (1918), whom set-theorists tell me influenced the development? Cf. my The Work of Kurt Gödel, p. 772.

Chapters 3, p. 7, line 7: Cantor's set theory as axiomatized with the aim of stopping short of the paradoxes.

P. 8. Is it worth remarking how trivially the two definitions of consistency are equivalent? For one direction, by observing that we are surely only interested in systems in which "0=0" is provable. And for the other, that we have the schema "A⊃(¬A⊃B)", whereby from any contradiction, every other formula (for example "0 ≠ 0") follows (both in classical and intuitionistic logic).

P. 12, last line, misprint: "0=0".

P. 27, end upper paragraph. At Königsberg on 7 September 1930, Gödel announced only his (first) incompleteness theorem, not his second one on the impossibility of a self-proof of consistency. Read Wang, J.S.L., vol. 46 (1981), top p. 655 (or indeed the Diskussion, not the later Nachtrag, in Erkenntnis 2). Of course, the first incompleteness theorem implies that a consistency proof, if it could be given for a system including number theory, would be limited in its scope to an incomplete system. However, in your context (p. 27), what you say could be misread as saying the second theorem was announced.

-3-

Chapter 4, p. 2. The top (fractional) paragraph could give the false impression that Church's December 1933 address <u>The Richard Paradox</u> antedated the realization and announcement to Church by Rosser that the Richard paradox could quite likely be worked in Church's system. In general, you should rethink your treatment to be sure that this transposition of events does not affect it more extensively. (I am commenting on some places I have noticed.) Sorry, I led you into this mistake by my July 1977 letter.

P. 2, last sentence of middle ⁋. As already explained, Church had proposed his thesis with λ-definability (it had gone beyond speculation, though of course not to publication) before Gödel introduced general recursive functions. (I think I found a place in your Chapter 4 or maybe 5 that seems to say so. At any rate, we now have Church's letter of November 29, 1935).

As I have checked from a book put out by the <u>Institute for Advanced Study</u>, Gödel came to Princeton in the fall of 1933 (contrary to your line 10 f.b.), though how much discussion took place between Church and Gödel before Gödel's spring-term-of-1934 lectures I don't know. Gödel introduced the general recursive functions late in the lectures (they are in the last of nine sections of the notes) - for a guess, in April 1934.

P. 10, line 3 of lower ⁋. Maybe it's just a quibble. The idea of defining 1, 2, 3, ... (bottom of the page) in the λ-notation could be thought of as having had "lasting importance". Well, actually you haven't said it doesn't. Maybe you could rephrase things more positively.

P. 12 top. In line 2, "summer 1933" should read "June 1933". I put my thesis in the hands of the reading committee in June 1933 (mid-June I think), and left for Maine, returning in September for my final public oral examination. The next statement "In January 1934 ... came to suspect..." is of course definitely dated incorrectly. The fourth line is not quite accurate; rather than having "just about proved", we were at that stage just about convinced our plan for the proof would work. You may notice from the Chronology that my <u>Proof by Cases...</u> and <u>A Theory of P_0s. Int...</u> were resubmitted in revised versions in March and June 1934, respectively, and K-R's <u>Inconsistency...</u> not until November. The revised versions of the first two were used in substantial measure to support the details of K-R. Just at what stage R and I had "just about proved" the consistency is hard to say at this distance in time, and also because "just about proved" is a bit vague.

P. 15, 1 line before display. Typographical error: for " ɩ " read " ι ".

P. 13 last full ⁋. Since the definition of <u>normal form</u> was Church's (cf. page 14 middle), let's change "Kleene's" to "the" and maybe add at the end of the sentence "which Kleene took over from Church's lectures in the fall of 1931" (with reference?). Incidentally, my 1934 paper establishing proof, and definition, by cases (far from trivial) was crucial to the later papers.

P. 16, last full sentence is subtilely inaccurate. Is my remark in

-4-

Reminiscences of Logicians likewise inaccurate? The truth is that Church's choice of his system of integers was designed to facilitate recursive definitions ("definitions by induction"), and my substitute cut me off from that way of doing them. I didn't have a proof that there was no other way of doing them with my substitute, but no other way of doing them came to my mind. I didn't embark on a research project to try for recursive definitions with my new system of positive integers. I immediately went back to Church's integers, and then got the predecessor function with them. Thirty-one years later, Dana Scott used another system of positive integers (maybe the same as my 1932 one) and got recursive definitions with it. Compare A.H.C., 1981, p. 56, low in rt. col., where the facts are stated more fully than in Reminiscences of Logicians.

P. 18 top sentence. Credit for substantially launching the idea is due to Church (cf. the enclosed letter, last sentence of ¶1). I was not on the scene in Princeton at the time Rosser's scheme for proving inconsistency made a rethinking of my thesis necessary, and I do think I was not entirely unaware, independently of Church, that a theory of definitions of functions was present, and of interest, separately from the logical structure of Church's system. I ventured to say so in A.H.C. vol. 3 (1981), p. 57, left col. lines 13-8 f.b., implying that the credit should not quite all of it be Church's. And indeed an immense amount of space in my unrevised thesis is devoted to studying what functions are "definable (formally)" (quote from p. 23). But your text seems to give the credit for the launching all to me (as I suppose, the paper does).

Pp. 20, 21. The reference to "every function recursive in the limited sense of Gödel 1931" and the footnote mention of the constant and identity functions were not in my October 1933 MSS. Though, of course, I had studied Gödel's 1931 paper, there is no reference to Gödel in that MSS. Before Gödel's 1934 lectures, we had thought of the [primitive] recursive functions as those definable using [primitive] recursion repeatedly in any order ((2) on Davis p. 14), substitution (Gödel's word on Davis p. 15 line 4) or explicit definitions (my IM 1952 p. 220), and the constant & successor functions. Wasn't this exact enough? Of course, the treatment became much more elegant when Gödel, using his identity functions, reduced the need for substitutions to just those of the form (IV) (IM p. 219).

P. 23, top line. I first learned of Ackermann's function from Gödel's 1934 lectures.

P. 33, beginning line 3 f.b. Surely this must be rewritten to make it clear that Church's Dec. 30, 1933 address, was a sequel to the communication by Rosser to Church of R's plan to derive the Richard paradox in Church's system (albeit a plan not fully carried out till considerably later).

P. 34, last paragraph. No! The published text of my A Theory of Positive Integers..., which you are consulting, is the revised text, received by Amer. Jour. of Math. June 18, 1934. Revised deliberately to include what Rosser and I would use in our proof of the inconsistency of Church's system. Section 19, "A representation of the logic C_1 within itself", and the theory of metads, was none of it in the unrevised version submitted for publication in October 1933 (in particular not the formulas ʊ and F). Of course, for your purpose, it probably comes to more or less the same that Church, while writing The Richard Paradox, already knew Rosser's plan for deriving the contradiction in his systems. I now think the word "retain" in my July 1977 letter page 2 line 4 of the middle paragraph, and in A.H.C. 1981 p. 57 left col. line 21 f.b., is misleading;

-5-

"contain" would be better. The revision of my thesis was not just by omitting
⌐(1) and (2),⌐things to get⌐but also by adding some things under (1) and (2),
and modifying the formulations of others. I did not spot any other place
you assumed that, because something is in my 1935 paper as published, it was
in the MSS of October 1933; but that assumption could be risky.

      I am very sorry I misled you in my letter of July 1977 (my false statement
which you repeat on p. 35 lines 2-4).

      On your p. 35, lines 5-6 there is a subtile distortion of the last clause
on p. 1 of my July 1977 letter, which I think is correct. It is the immediately
preceding clause which is false (the truth: Rosser suspected the inconsistency
in the fall of 1933 and told Church; I was informed of his suspicions in January
of 1934 or maybe December 1933). Consider four statements:

(a)   perhaps by the end of March or April we had pretty well confirmed the fact.

(b)   the actual proof of inconsistency was not completed prior to about April 1934.

(c)   the actual proof of inconsistency was completed about April 1934.

(d)   by about April 1934 Rosser and I were pretty sure the actual proof of
      inconsistency could be completed.

Statement (a) is as in my letter. It seems to me that your p. 35 lines 5-6 is
equivalent to the conjunction of (b) and (c) . Now (b) is true but (c) is
false, while (d) is probably true (having omitted "perhaps" from (a) and
replaced "end of March or April" by simply "April").

      Of course -- didn't I say it earlier? -- the date by which we were pretty
sure is pretty vague. As to the date we actually had the actual proof (all
written out and double-checked), it is presumably beyond recovery now. I rewrote
my two previous papers P. by C. and T. of P.I. successively, to provide the
foundation for The inconsistency.... I presumably didn't start writing The In.
till I had those finished (June 1934), though I surely had the plan. The In.
wasn't submitted till Nov. 1934, though the part relating to Church's system
was no doubt finished before we tacked on §4 concerning a system of Curry.
(On p. 13 line 11, is it quite accurate then to say "at the same time"? Of
course it was published at the same time.)

      I think the confusion I caused by my error in July 1977 affects the rest of
p. 35 and much of p. 36.

      P. 36, line 3 f.b. Because I only returned to Princeton on February 7, 1934,
and it antedates Gödel's introduction of general recursive functions in about
April, 1934, your "around the spring of 1934" has to mean February or March
or improbably very early April) 1934, and spring begins April 21. Of course,
the spring term begins in January or February.

-6-

P. 37, line 2. Church sent in a preliminary report of his paper An Unsolvable Problem..., received March 22, 1935. The paper itself was not submitted for publication before July 1935 at the earliest, as follows from (i) its references to my λ-def. & Rec. and Gen. Rec. Fcns...., which I only completed (giving him copies) at the end of June 1935 (sending in the two papers and their abstracts simultaneously), and (ii) the quote from his letter of 6 July 1935 saying he had "made a number of changes, since he saw [me]" in his paper, An Unsolvable Problem..... He saw me late in June.

Chapter 5, pp. 1 and 5. The terminology "primitive recursive functions" was very consciously introduced by me in my paper General Recursive Functions... (last line of p. 727 and top of p. 729 (Davis, pp. 237, 239)). Péter 1934 on p. 613 has "primitive Rekursion" but not "primitive rekursive Funktion". Also she has 'Die eingeschackelte Rekursion", "Die mehrfache Rekursion", "Die Wertverlaufsrekursion", etc. Cf. my Bull. (n.s.) Amer. Math. Soc., vol. 5 (1981), Fn. 4.

P. 3, lines 4 and 5, read "calculable" for "solvable".

Although you don't cite it, I assume you are aware of van Heijenoort, the Gödel quote on p. 619.

P. 23, line 6, better "the" or "his" than "a". Cf. A.H.C., vol. 3 (1981), p. 60, left col.

P. 41, top line. In this connection my attention has been called to Jill Humphries' article, Gödel's Proof and the Liar Paradox, Notre Dame J. of Formal Logic, Vol. 20, (1979), pp. 535-544. (I haven't studied it closely.)

In connection with p. 43, iii) and the sentence beginning p. 43 bottom line, I am puzzled by your saying I relied on Theorem XVII. Have you noticed the Addendum (privately circulated in 1936) published with the Davis reprint p. 253? On reexamining that addendum today, it appears "$(y) \underline{T}_1(\theta(\underline{b}), \theta(\underline{b}), y)$" should instead read "$(\underline{Ey})\underline{T}(\underline{f}, \underline{f}, y)$, i.e. $(\underline{Ey})\underline{T}_1(\theta(\underline{q}), \theta(\underline{q}), \overline{y})$".

I have found your five chapters very interesting.

Sincerely yours,

Stephen C. Kleene

SCK/dcm

Enclosures: letter of Church to Kleene, September 29, 1935.
      Chronology, by Kleene, 22 October 1981 .
Under separate cover: Reprints of my 1981 articles in Annals of the History of Computing, vol. 3 and Bulletin (n.s.) of the American Mathematical Society, vol. 5.

-7-

P.S.   Regarding the middle paragraph of page 4 of this letter, I now realize
that Gödel in 1931 in using "substitution" without admitting the identity functions
initially slipped a bit, whereas if one is thinking in terms of "explicit definition"
generally (IM p. 220) the identity functions enter through the trivial cases that the
defining expression for the ambiguous value is just one of the independent variables.

In the October 1933 MSS of my thesis, I did generally what I describe in A.H.C.
1981 from p. 57 left col. bottom   through p. 58 (but double recursion, p. 58 ¶ 3,
was only in the June 1934 revision).   For example (p. 58 ¶ 3) in October 1933 I
treated course-of-values recursion (with or without a parameter) before Péter 1934
appeared.   Although no notice was taken then of the constant and identity functions,
any one who had read the thesis would surely have been able to produce λ-definitions
in minutes.   I did not then relate my results to any previously defined class of
effectively calculable functions.   You are right that the tie in with Gödel's 1931
recursive functions was only introduced in the June 1934 revision.   In October
1933, I just dealt with every specific effectively calculable function or
effective-calculability-preserving operation we thought of (A.C.M. p. 57).

                                                    S. C. K.

CHRONOLOGY

S. C. Kleene
October 22, 1981

8 May 1933   K's Proof by cases ... rec'd by Annals.

Sept 1933   K's Theory of pos. int.... accepted for a PhD thesis by Princeton.
Kleene then went to Maine.

9 Oct 1933   K's Theory of pos. int. ... rec'd by Am. Jour.

Fall 1933   Rosser suspected contrad. in Church's system, and told Church.
Church spoke out on the significance of λ-def.
Gödel came to I.A.S.

30 Dec 1933   C's Richard paradox delivered as an address to M.A.A.

7 Feb 1934   **Kleene returned to Princeton.**

28 Mar 1934   Rev. MSS of K's Proof by cases ... rec'd by Annals.

c. Apr 1934   Gödel introduces (gen.) rec. fcns. in his I.A.S. lectures, §9 of notes.

18 Jun 1934   Rev. MSS of K's Theory of pos. int. ... rec'd by Am. Jour.

13 Nov 1934   K-R's Inconsistency rec'd by Annals.

22 Mar 1935   Abstract of C's Unsolvable problem ... (prelim. report) rec'd by A.M.S.
Abstract of C-R's Some properties of conversion rec'd by A.M.S.

26 Mar 1935   C's Freedom from contradiction communicated to N.A.S.

19 Apr 1935   Church speaks to A.M.S. on Unsolvable problem ... (prelim. report).

May 1935   Abstract of C's Unsolvable problem ... (prelim. report) pub'l in Bulletin.

1 July 1935   Abstract of K's λ-def. and rec. rec'd by A.M.S.
Abstract of K's Gen. rec. fcns. rec'd by A.M.S.
K's λ-def. and rec. rec'd by Duke Math. Jour.

6 July 1935   Church wrote Kleene, "I have made a number of changes, since I saw you, in my paper 'An unsolvable problem of elementary number theory'."

7 July 1935   K's Gen. rec. fcns. rec'd by Math. Ann.

Dear Mr. Adams:

Thank you for sending me a copy of the handwritten MSS of your Chapters 1-5.

In writing my two 1981 papers (reprints of which I am sending under separate cover) and my "Biographical Memoir" on Kurt Gödel for the National Academy of Sciences (about to be sent to press), I have come up with some more details about the events under discussion. So I have reread my letters to you of July 28, 1977 and July 17, 1979 to see what I would now change or amplify.

July 1979, third paragraph. Yes, Gödel was "study[ing] the problem of proving the consistency of analysis [and thus of arithmetic]" when he obtained his famous 1931 results, as indicated by Wang in *Jour. Symbolic Logic,* vol. 46 (1981), pp. 654-655. And Gödel did know of Finsler's 1926 paper, according to Kreisel in *Zentralblatt,* vol. 401 (1979), p. 13.

July 1977, second ¶ on p. 2. "February 1934" is a misprint for "February 1932", as you must have realized. The history related in the last paragraph of p. 1 is more accurately as follows.

Rosser came up with the idea of proving the inconsistency of Church's system (with an outline of the reasoning to be used) in the fall of 1933, but of course without knowing instantly whether or not his proposed reasoning could be carried out in full detail. He told his proposed plan for proving the inconsistency to Church, who reacted in two ways. First, he spoke out on the significance of λ-definability as a notion of number theory of interest in its own right, which he "afterwards" communicated "to [me] in connection with the consequent revision of [my] thesis." I am quoting from his letter to me of November 29, 1935 (copy enclosed). I was off in Maine, so I'm not sure just how early I learned about all this. I remember that in December 1933 or January

1934, Rosser wrote me his plan for proving the inconsistency of Church's system and recruited me to join him in carrying it out. Of course, carrying out the proof called for expertise in developing theory within Church's system, and this is just what I had. I remember working on it, some weeks, in my farmhouse in Maine, and then being offered a Research Assistantship to continue the work at Princeton, where I could work more efficiently. Of course, in Church's letter "Rosser's discovery of the contradiction in [his] system" means the discovery of the idea for proving a contradiction in the system. It took some time to actually formalize Rosser's sketch, and our confidence that the proof could be done no doubt gradually increased as I ran into no snags in my research on it. We were probably pretty sure the proof would go through "by end of March or April", as I wrote. Church's other reaction to learning of Rosser's plan in the fall of 1933 was to write, deliver and publish, his paper, The Richard Paradox. As stated above, I was in Maine while this was going on. My return to Princeton was on February 7, 1934. But Rosser told then very explicitly that writing The Richard Paradox was a response by Church to his [R's] proposal of his plan to derive a contradiction in Church's system.

I enclose a Chronology that I recently had another occasion to compose. The events described in my Annals of the History of Computing, vol. 3 (1981) paper on p. 59 left column lines 4-1 f.b. and right column lines 8-14 (in the second paragraph) belong in the period shown by the box in my Chronology.

Evidently, when I wrote you in July 1977, I had forgotten both about Church's letter of November 29, 1936, and about Church's paper "The Richard Paradox" (which had no effect on my research) with its December 30, 1933 date of delivery to the A.M.S., both of which put Rosser's discovery of his plan well back into the fall of 1933. It was easier for me to forget these, since I was not in Princeton when the events took place.

Now I will comment on points in your MSS.

Chapter 1, p. 17. You cite Zemelo (1908) and Fraenkel (1922). Should you also mention Skolem (1922) and Weyl (1910), (1918), whom set-theorists tell me influenced the development? Cf. my "The Work of Kurt Gödel", p. 772.

Chapters 3, p. 7, line 7: Cantor's set theory as axiomatized with the aim of stopping short of the paradoxes.

P. 8. Is it worth remarking how trivially the two definitions of consistency are equivalent? For one direction, by observing that we are surely only interested in systems in which "$0 = 0$" is provable. And for the other, that we have the schema "$A \supset (\neg A \supset B)$", whereby from any contradiction, every other formula (for example "$0 \neq 0$") follows (both in classical and intutionistic logic).

P. 12, last line, misprint: "$0 = 0$".

P. 27, end upper paragraph. At Königsberg on 7 September 1930, Gödel announced only his (first) incompleteness theorem, not his second one on the impossibility of a self-proof of consistency. Read Wang, *J.S.L.*, vol. 46 (1981), top p. 655 (or indeed the "Diskussion," not the later "Nachtrag," in *Erkenntnis* 2). Of course, the first incompleteness theorem implies that a consistency proof, if it could be given for a system including number theory, would be limited in its scope to an *incomplete* system. However, in your context (p. 27), what you say could be misread as saying the second theorem was announced.

Chapter 4, p. 2. The top (fractional) paragraph could give the false impression that Church's December 1933 address "The Richard Paradox" antedated the realization and announcement to Church by Rosser that the Richard paradox could quite likely be worked in Church's system. In general, you should rethink your treatment to be sure that this transposition of events does not affect it more extensively. (I am commenting on some places I have noticed.) Sorry, I led you into this mistake by my July 1977 letter.

P. 2, last sentence of middle ¶. As already explained, Church had proposed his thesis with λ-definability (it had gone beyond speculation, though of course not to publication) before Godel introduced general recursive functions. (I think I found a place in your Chapter 4 or maybe 5 that seems to say so. At any rate, we now have Church's letter of November 29, 1935).

As I have checked from a book put out by the Institute for Advanced Study, Gödel came to Princeton in the fall of 1933 (contrary to your line 10 f.b.), though how much discussion took place between Church and Gödel before Gödel's spring-term-of-1934 lectures I don't know. Gödel introduced the general recursive functions late in the lectures (they are in the last of nine sections of the notes) – for a guess, in April 1934.

P. 10, line 3 of lower ¶. Maybe it's just a quibble. The idea of defining 1, 2, 3, ... (bottom of the page) in the λ-notation could be thought of as having had "lasting importance". Well, actually you haven't said it doesn't. Maybe you could rephrase things more positively.

P. 12 top. In line 2, "summer 1933" should read "June 1933". I put my thesis in the hands of the reading committee in June 1933 (mid-June I think), and left for Maine, returning in September for my final public oral examination. The next statement "In January 1934 ... came to suspect ..." is of course definitely dated incorrectly. The fourth line is not quite accurate; rather than having "just about proved", we were at that stage just about convinced our plan for the proof would work. You may notice from the Chronology that my "Proof by Cases" and "A Theory of Pos. Int..." were resubmitted in revised versions in March and June 1934, respectively, and K-R's "Inconsistency ..." not until November. The revised versions of the first two were used in substantial measure to support the details of K-R. Just at what stage R and I had "just about proved" the consistency is hard to say at this distance in time, and also because "just about proved" is a bit vague.

P. 15, 1 line before display. Typographical error: for "i" read " i".

P. 13 last full ¶. Since the definition of *normal form* was Church's (cf. page 14 middle), let's change "Kleene's" to "the" and maybe add at the end of the sentence "which Kleene took over from Church's lectures in the fall of 1931" (with reference?). Incidentally, my 1934 paper establishing proof, and definition, by cases (far from trivial) was crucial to the later papers.

P. 16, last full sentence is subtilely [sic] inaccurate. Is my remark in "Reminiscences of Logicians" likewise inaccurate? The truth is that Church's choice of his system of integers was designed to facilitate recursive definitions ("definitions by induction"), and my substitute cut me off from that way of doing them. I didn't have a proof that there was no other way of doing them with my substitute, but no other way of doing them came to my mind. I didn't embark on a research project to try for recursive definitions with my new system of positive integers. I immediately went back to Church's integers, and then got the predecessor function with them. Thirty-one years later, Dana Scott used another system of positive integers (maybe the same as my 1932 one) and got recursive definitions with it. Compare A.H.C., 1981, p. 56, low in rt. col. , where the facts are stated more fully than in "Reminiscences of Logicians."

P. 18 top sentence. Credit for substantially launching the idea is due to Church (cf. the enclosed letter, last sentence of ¶ 1). I was not on the scene in Princeton at the time Rosser's scheme for proving inconsistency made a rethinking of my thesis necessary, and I do think I was not entirely unaware, independently of Church, that a theory of definitions of functions was present, and of interest, separately from the logical structure of Church's system. I ventured to say so in *A.H.C.* vol. 3 (1981), p. 57, left col. lines 13–8 f.b., implying that the credit should not quite all of it be Church's. And indeed an immense amount of space in my unrevised thesis is devoted to studying what functions are "definable (formally)" (quote from p. 23). But your text seems to give the credit for the launching all to me (as I suppose, the paper does).

Pp. 20, 21. The reference to "every function recursive in the limited sense of Gödel 1931" and the footnote mention of the constant and identity functions were not in my October 1933 MSS. Though, of course, I had studied Gödel's 1931 paper, there is no reference to Gödel in that MSS. Before Gödel's 1934 lectures, we had thought of the [primitive] recursive functions as those definable using [primitive] recursion repeatedly in any order ((2) on Davis p. 14), substitution (Gödel's word on Davis p. 15 line 4) or explicit definitions (my IM 1952 p. 220), and the constant & successor functions. Wasn't this exact enough? Of course, the treatment became much more elegant when Gödel, using his identity functions, reduced the need for substitutions to just those of the form (IV) (IM p. 219).

P. 23, top line. I first learned of Ackermann's function from Gödel's 1934 lectures.

P. 33, beginning line 3 f.b. Surely this must be rewritten to make it clear that Church's Dec. 30, 1933 address, was a sequel to the communication by Rosser to Church of R's plan to derive the Richard paradox in Church's system (albeit a plan not fully carried out till considerably later).

P. 34, last paragraph. No! The published text of my "A Theory of Positive Integers ...", which you are consulting, is the revised text, received by *Amer. Jour. of Math.* June 18, 1934. Revised deliberately to include what Rosser and I would use in our proof of the inconsistency of Church's system. Section 19, "A representation of the logic C_1 within itself", and the theory of metads, was none of it in the unrevised version submitted for publication in October 1933 (in particular not the formulas $\mathfrak{U}$ and F). Of course, for your purpose, it probably comes to more or less the same that Church, while writing "The Richard Paradox," already knew Rosser's plan for deriving the contradiction in his systems. I now think the word "retain" in my July 1977 letter page 2 line 4 of the middle paragraph, and in *A.H.C.* 1981 p. 57 left col. line 21 f.b., is misleading; "contain" would be better. The revision of my thesis was not just

by omitting things to get (1) and (2), but also by adding some things under
(1) and (2), and modifying the formulations of others. I did not spot any other
place you assumed that, because something is in my 1935 paper as published,
it was in the MSS of October 1933; but that assumption could be risky.

I am very sorry I misled you in my letter of July 1977 (my false statement
which you repeat on p. 35 lines 2–4).

On your p. 35, lines 5–6 there is a subtile [sic] distortion of the last clause
on p. 1 of my July 1977 letter, which I think is correct. It is the immediately
preceding clause which is false (the truth: Rosser suspected the inconsistency
in the fall of 1933 and told Church; I was informed of his suspicions in January
of 1934 or maybe December 1933). Consider four statements:

(a) perhaps by the end of March or April we had pretty well confirmed
the fact.
(b) the actual proof of inconsistency was not completed prior to about
April 1934.
(c) the actual proof of inconsistency was completed about April 1934.
(d) by about April 1934 Rosser and I were pretty sure the actual proof of
inconsistency could be completed.

Statement (a) is as in my letter. It seems to me that your p. 35 lines 5–6
is equivalent to the conjunction of (b) and (c). Now (b) is true but (c) is false,
while (d) is probably true (having omitted "perhaps" from (a) and replaced
"end of March or April" by simply "April").

Of course – didn't I say it earlier? – the date by which we were pretty sure
is pretty vague. As to the date we actually had the actual proof (all written
out and double-checked), it is presumably beyond recovery now. I rewrote my
two previous papers "P. by C." and "T. of P.I." successively, to provide the
foundation for "The inconsistency ..." I presumably didn't start writing "The

In." till I had those finished (June 1934), though I surely had the plan. The In. wasn't submitted till Nov. 1934, though the part relating to Church's system was no doubt finished before we tacked on § 4 concerning a system of Curry. (On p. 13 line 11, is it quite accurate then to say "at the same time"? Of course it was published at the same time.)

I think the confusion I caused by my error in July 1977 affects the rest of p. 35 and much of p. 36.

P 36, line 3 f.b. Because I only returned to Princeton on February 7, 1934, and it antedates Gödel's introduction of general recursive functions in about April, 1934, your "around the spring of 1934" has to mean February or March or improbably very early April) 1934, and spring begins April 21. Of course, the spring term begins in January or February.

P. 37, line 2. Church sent in a *preliminary report* of his paper "An Unsolvable Problem ..." , received March 22, 1935. The paper itself was not submitted for publication before July 1935 at the earliest, as follows from (i) its references to my "λ-def. & Rec." and "Gen. Rec. Fcns..." , which I only completed (giving him copies) at the end of June 1935 (sending in the two papers and their abstracts simultaneously), and (ii) the quote from his letter of 6 July 1935 saying he had "made a number of changes, since he saw [me]" in his paper, "An Unsolvable Problem..." He saw me late in June.

Chapter 5, pp. 1 and 5. The terminology "primitive recursive functions" was very consciously introduced by me in my paper "General Recursive Functions..." (last line of p. 727 and top of p. 729 (Davis, pp. 237, 239)). Péter 1934 on p. 613 has "primitive Rekursion" but not "primitive rekursive Funktion". Also she has "Die eingeschackelte Rekursion", "Die mehrfache Rekursion", "Die Wertverlaufsrekursion" , etc. Cf. my *Bull. (n.s.) Amer. Math. Soc.,* vol. 5 (1981), Fn. 4.

P. 3, lines 4 and 5, read "calculable" for "solvable".

Although you don't cite it, I assume you are aware of van Heijenoort, the Gödel quote on p. 619.

P. 23, line 6, better "the" or "his" than "a". Cf. *A.H.C.*, vol. 3 (1981), p. 60, left col.

P. 41, top line. In this connection my attention has been called to Jill Humphries' article, "Gödel's Proof and the Liar Paradox", *Notre Dame J. of Formal Logic,* Vol. 20, (1979), pp. 535-544. (I haven't studied it closely.)

In connection with p. 43, iii) and the sentence beginning p. 43 bottom line, I am puzzled by your saying I relied on Theorem XVII. Have you noticed the Addendum (privately circulated in 1936) published with the Davis reprint p. 253? On reexamining that addendum today, it appears "$(y)\underline{T}_1(\theta(\underline{b}), \theta(\underline{b}), \underline{y})$" should instead read "$(\underline{Ey})\underline{T}_1(\underline{f}, \underline{f}, \underline{y})$, i.e. $(\underline{Ey})\underline{T}_1(\theta(\underline{q}), \theta(\underline{q}), \underline{y})$".

I have found your five chapters very interesting.

Sincerely yours,

Stephen C. Kleene

SCK/dcm

Enclosures: letter of Church to Kleene, September 29, 1935.
Chronology, by Kleene, 22 October 1981.
Under separate cover: Reprints of my 1981 articles in *Annals of the History of Computing,* vol. 3 and *Bulletin (n.s.) of the American Mathematical Society,* vol. 5.

P.S. Regarding the middle paragraph of page 4 of this letter, I now realize that Gödel in 1931 in using "substitution" without admitting the identity functions initially slipped a bit, whereas if one is thinking in terms of "explicit definition" generally (IM p. 220) the identity functions enter through the trivial cases that the defining expression for the ambiguous value is just one of the independent variables.

In the October 1933 MSS of my thesis, I did generally what I describe in *A.H.C.* 1981 from p. 57 left col. bottom through p. 58 (but double recursion, p. 58 ¶ 3, was only in the June 1934 revision). For example (p. 58 ¶ 3) in October 1933 I treated course-of-values recursion (with or without a parameter) before Péter 1934 appeared. Although no notice was taken then of the constant and identity functions, anyone who had read the thesis would surely have been able to produce λ-definitions in minutes. I did not then relate my results to any previously defined class of effectively calculable functions. You are right that the tie in with Gödel's 1931 recursive functions was only introduced in the June 1934 revision. In October 1933, I just dealt with every specific effectively calculable function or effective-calculability-preserving operation we thought of (*A.C.M.* p. 57).   S. C. K.

S. C. Kleene

October 22, 1981

CHRONOLOGY

8 May 1933	K's "Proof by cases ..." rec'd by *Annals*.
Sept 1933	K's "Theory of pos. int. ..." accepted for a PhD thesis by Princeton.
	Kleene then went to Maine.
9 Oct 1933	K's "Theory of pos. int. ..." rec'd by *Am. Jour.*
Fall 1933	Rosser suspected contrad. in Church's system, and told Church.
	Church spoke out on the significance of λ-def.
	Gödel came to I.A.S.
30 Dec 1933	C's "Richard paradox" delivered as an address to M.A.A.
7 Feb 1934	**Kleene returned to Princeton.**

28 Mar 1934	Rev. MSS of K's "Proof by cases..." rec'd by *Annals*.
c. Apr 1934	Gödel introduces (gen.) rec. fcns. in his I.A.S. lectures, § 9 of notes.
18 Jun 1934	Rev. MSS of K's "Theory of pos. int. ..." rec'd by *Am. Jour.*
13 Nov 1934	K-R's "Inconsistency" rec'd by *Annals*.
22 Mar 1935	Abstract of C's "Unsolvable problem ... (prelim. report)" rec'd by A.M.S.
	Abstract of C-R's "Some properties of conversion" rec'd by A.M.S.
26 Mar 1935	C's "Freedom from contradiction" communicated to N.A.S.
19 Apr 1935	Church speaks to A.M.S. on "Unsolvable problem ... (prelim. report)"
May 1935	Abstract of C's "Unsolvable problem ... (prelim. report)" pub'l in *Bulletin*
1 July 1935	Abstract of K's "λ-def. and rec." rec'd by A.M.S.
	Abstract of K's "Gen rec. fcns." rec'd by A.M.S.
	K's "λ-def. and rec." rec'd by *Duke Math. Jour.*
6 July 1935	Church wrote Kleene "I have made a number of changes, since I saw you, in my paper 'An unsolvable problem of elementary number theory'."
7 July 1935	K's "Gen. rec. fcns." rec'd by *Math. Ann.*

B.8. Stephen C.Kleene to Author, November 13, 1981

Mathematics Department
University of Wisconsin—Madison

UW madison

Van Vleck Hall
480 Lincoln Drive
Madison, Wisconsin 53706
Telephone: (608) 263-3053

November 13, 1981

Mr. R. G. Adams,
St. Albans College,
29 Hatfield Road,
St. Albans, Herts.,
ENGLAND

Dear Mr. Adams: p. 27,

   <u>Chapter 3, end upper paragraph.</u> A further reference to the effect
that Gödel didn't announce his second theorem at Königsberg on 7 Sept. 1930 is Gödel in
<u>Zentralblatt,</u> vol. 2 (1932), p. 322, the last two sentences. (I was also inaccurate.
In AHC vol.3 (1981) p. 52 rt. col. line 5, add at the end "the first of".)

   <u>Chapter 4, bottom line p. 20, and my Nov. 3 p. 7 top ¶.</u> Was the definition of
[primitive] recursive function in Gödel's 1931 paper actually defective?

   Of course, $\underline{x+y}$ is recursive (Gödel p. 16 Fn. 31 and p. 17, in Davis 1965); if it wasn't,
Gödel's definition would be no good. The usual recursion for $\underline{x+y}$ (stated with the recursion
on the first variable, to match Gödel's form of the recursion schema, (2) p. 14) is

$$\begin{cases} \phi(0,\underline{y}) = \underline{y}, \\ \phi(\underline{x+1},\underline{y}) = \phi(\underline{x},\underline{y}) + 1. \end{cases}$$

Somehow, Gödel had to be satisfied that this, or some equivalent of it, was allowed under
his definition. The first equation fits his schema (2) when it is considered as $\phi(0,\underline{y}) = \psi(\underline{y})$
where ψ is the identity function $U_1^1(\underline{y}) = \underline{y}$.

   Now I observe that all the identity functions $U_{\underline{i}}^{\underline{n}}$ are recursive under Gödel's 1931
definition as he intended it to be understood, if we make the following two assumptions.
(A) Though he doesn't explicitly state it, do you believe he didn't intend to allow
recursion to hold for the case $\underline{n} = 1$, with his equations (2) becoming

(2_1)
$$\begin{cases} \phi(0) = \underline{c}, \\ \phi(\underline{k+1}) = \mu(\underline{k},\phi(\underline{k})) \end{cases}$$

with $\underline{c}$ a constant? How otherwise does he propose to define α on p. 16 (cf. the recursion
for α actually displayed on p. 44) or $\underline{x}!$ p. 44? True, the last is in 1934; but he is
apparently operating the same way there. (B) I construe the reference to (2) in Fn. 27
p. 15 (or on p. 43 just after (2)) to mean that the $\mu(\underline{x}_1,\underline{x}_2,\ldots,\underline{x}_{n+1})$ for the second
equation can be replaced by a function of any non-empty proper sublist of $\underline{x}_1,\underline{x}_2,\ldots,\underline{x}_{n+1}$
(the empty sublist is provided for anyway by using a constant function as the μ), and
similarly the $\psi(\underline{x}_1,\ldots,\underline{x}_{n-1})$ of the first equation. It is so familiar that, e.g., in the
induction step of a recursion $\phi(\underline{k+1},\underline{x}_2,\ldots,\underline{x}_n)$ is allowed to depend on each of
$\underline{k}, \phi(\underline{k},\underline{x}_2,\ldots,\underline{x}_n), \underline{x}_2, \ldots, \underline{x}_n$, but in particular cases need not depend on all $\underline{n}+1$ of them.
For example, in the familiar recursion for $\underline{x+y}$, neither the $\underline{k}$ nor the $\underline{x}_2$ are used (as I took
it above on $\underline{x}$); for $\underline{x \cdot y}$, the $\underline{k}$ is not used. In particular, I assume he intended (2_1) to
include the cases of simply $\mu(\underline{k})$ or simply $\mu(\phi(\underline{k}))$ on the right of the second equation.

   Under assumptions (A) and (B), I now define all $U_{\underline{i}}^{\underline{n}}$.

$$\phi_1(\underline{x}) = \underline{x+1}.$$
$$\begin{cases} \phi_2(0) = 0, \\ \phi_2(\underline{k+1}) = \phi_1\,\phi_2(\underline{k})). \end{cases}$$

- 2 -

Then, for all $\underline{x}, \phi_2(\underline{x}) = \underline{x}$; i.e. $\phi_2 = \underline{U}_1^1$. Using substitution (with Gödel's allowance in Fn. 27 that not all the variables on the left have to occur on the right), for any $\underline{n} > 1$ and $1 \le \underline{i} \le \underline{n}$,

$$\underline{U}_{\underline{i}}^{\underline{n}}(\underline{x}_1, \ldots, \underline{x}_{\underline{i}}, \ldots, \underline{x}_{\underline{n}}) = \underline{U}_1^1(\underline{U}_1^1(x_{\underline{i}})).$$

So your statment (bottom p. 20) that "a third set of initial functions was needed" can be questioned. Of course, the introduction of the $\underline{U}_{\underline{i}}^{\underline{n}}$ as the third set makes the treatment neater in that it removes the need for flexibility in the use of the recursion and substitution schemas (Davis p. 43 Fn. 2 and my IM p. 221 next to last ¶).

<u>Chapter 4, relations between my 1935 and Gödel.</u> In connection with your top p. 22 and my p. 224,top, it may be mentioned that through the λ-definition of the least-number operator I had a more powerful method than is available in primitive recursion theory , which therefore may be easier to use for some applications than Gödel's methods.

Undoubtedly Fn. ‡ on p. 219 was added after I heard the beginnings of Gödel's 1934 lectures (I didn't think of the identity functions before Gödel brought them in), and likewise the passage beginning at the bottom of p. 223 with its three footnotes. I have already acknowledged first hearing of Ackermann's function from Gödel (used also in the last ¶ of §15). in my 1935 paper
I think, however, the passages just cited are the only ones directly related to Gödel. Of course, the idea of self-representation of a formalism comes from Gödel 1931; so the plan Rosser and I had for proving Church's system inconsistent owes something to Gödel. But my particular method of representation, using the metads (§ 19), is different from Gödel's method of numbering. If my recollection is correct, I started out with a Gödel-like numbering, and then decided that in my context the metads were available and easier to use.

Of course, I drew heavily on Gödel's methods in my <u>Gen. Rec. Fcns.</u> (1936). The use of Gödel numbering there is, I think, the third time it was used in the literature (after Gödel's use, and Tarski's in his work on the truth concept, if we don't count my metads). The Gödel numbering method is somewhat changed in my IM § 50 (influenced by Hilbert and Bernays 1939). In IM § 45 (e.g. beginning at the bottom line of p. 227) I use Gödel's methods.

Sincerely yours,

Stephen C. Kleene

Dear Mr. Adams:

Chapter 3, p. 27, end upper paragraph. A further reference to the effect that Gödel didn't announce his second theorem at Königsberg on 7 Sept. 1930 is Gödel in *Zentralblatt,* vol. 2 (1932), p. 322, the last two sentences. (I was also inaccurate. In *ARC* vol. 3 (1981) p. 52 rt. col. line 5, add at the end "the first of".)

Chapter 4, bottom line p. 20, and my Nov. 3 p. 7 top ¶. Was the definition of [primitive] recursive function in Gödel's 1931 paper actually defective?

Of course, $\underline{x} + \underline{y}$ is recursive (Gödel p. 16 Fn. 31 and p. 17, in Davis 1965); if it wasn't, Gödel's definition-would be no good. The usual recursion for $\underline{x} + \underline{y}$ (stated with the recursion on the first variable, to match Gödel's form of the recursion schema, (2) p. 14) is

$$\begin{cases} \phi(0, \underline{y}) = \underline{y}, \\ \phi(\underline{x} + 1, \underline{y}) = \phi(\underline{x}, \underline{y}) + 1. \end{cases}$$

Somehow, Gödel had to be satisfied that this, or some equivalent of it, was allowed under his definition. The first equation fits his schema (2) when it is considered as $\phi(0, \underline{y}) = \psi(\underline{y})$ where ψ is the identity function $\underline{U}_1^1(\underline{y}) = \underline{y}$.

Now I observe that all the identity functions $\underline{U}_j^n$ are recursive under Gödel's 1931 definition as he intended it to be understood, if we make the following two assumptions.

(A) Though he doesn't explicitly state it, do you believe he didn't intend to allow recursion to hold for the case $\underline{n} = 1$, with his equations (2) becoming

(2_1)
$$\begin{cases} \phi(0) = \underline{c}, \\ \phi(\underline{k} + 1) = \mu(\underline{k}, \phi(\underline{k})) \end{cases}$$

with c a constant? How otherwise does he propose to define α on p. 16 (cf. the recursion for α actually displayed on p. 44) or $x!$ p. 44? True, the last is in 1934; but he is apparently operating the same way there.

(B) I construe the reference to (2) in Fn. 27 p. 15 (or on p. 43 just after (2)) to mean that the $\mu((x_1, \ldots, x_{n+1}))$ for the second equation can be replaced by a function of any non-empty proper sublist of $(x_1, \ldots, x_{n+1})$ (the empty sublist is provided for anyway by using a constant function as the μ), and similarly the $\psi((x_1, \ldots, x_{n-1}))$ of the first equation.

It is so familiar that, e.g., in the induction step of a recursion $\phi(k + 1, x_2, \ldots, x_n)$ is allowed to depend on each of k, $\phi(k, x_2, \ldots, x_n), x_2, \ldots, x_n$ but in particular cases need not depend on all $n + 1$ of them.

For example, in the familiar recursion for $x + y$, neither the k nor the x_2 are used (as I took it above on x); for $x \cdot y$, the k is not used. In particular, I assume he intended (2_1) to include the cases of simply $\mu(k)$ or simply $\mu(\phi(k))$ on the right of the second equation.

Under assumptions (A) and (B), I now define all U_j^n.

$$\phi_1(x) = x + 1.$$

$$\left\{ \begin{array}{l} \phi_2(0) = 0, \\ \phi_2(k + 1) = \phi_1(\phi_2(k)). \end{array} \right. \quad \centerdot$$

Then, for all x, $\phi_2(x) = x$; i.e. $\phi_2 = U_1^1$. Using substitution (with Gödel's allowance in Fn. 27 that not all the variables on the left have to occur on the right), for any $n > 1$ and $1 \leq j \leq n$,

$$U_j^n(x_1, \ldots, x_j, \ldots, x_n) = U_1^1(U_1^1(x_j)).$$

So your statement (bottom p. 20) that "a third set of initial functions was needed" can be questioned.

Of course, the introduction of the $\underline{U}^n_j$ the third set makes the treatment neater in that it removes the need for flexibility in the use of the recursion and substitution schemas (Davis p. 43 Fn. 2 and my IM p. 221 next to last ¶).

Chapter 4, relations between my 1935 and Gödel. In connection with your top p. 22 and my p. 224 top, it may be mentioned that through the λ-definition of the least-number operator I had a more powerful method than is available in primitive recursion theory, which therefore may be easier to use for some applications than Gödel's methods.

Undoubtedly Fn. ‡; on p. 219 was added after I heard the beginnings of Gödel's 1934 lectures (*I* didn't think of the identity functions before Gödel brought them in), and likewise the passage beginning at the bottom of p. 223 with its three footnotes. I have already acknowledged first hearing of Ackermann's function from Gödel (used also in the last ¶ of § 15).

I think, however, the passages just cited are the only ones in my 1935 paper directly related to Gödel. Of course, the idea of self-representation of a formalism comes from Gödel 1931; so the plan Rosser and I had for proving Church's system inconsistent owes something to Gödel. But my particular method of representation, using the metads (§ 19) is different from Gödel's method of numbering. If my recollection is correct, I started out with a Gödel-like numbering, and then decided that in my context the metads were available and easier to use.

Of course, I drew heavily on Gödel's methods in my "Gen. Rec. Fcns."
(1936). The use of Gödel numbering there is, I think, the third time it was used
in the literature (after Gödel's use, and Tarski's in his work on the truth concept,
if we don't count my metads). The Gödel numbering method is somewhat
changed in my 1M § 50 (influenced by Hilbert and Bernays 1939). In IM § 45
(e.g. beginning at the bottom line of p. 227) I use Gödel's methods.

Sincerely yours,

Stephen C. Kleene

B.9. J. Barkley Rosser to Author, November 19, 1981

J. BARKLEY ROSSER
UNIVERSITY OF WISCONSIN–MADISON
MADISON 53706

Nov. 19, 1981

MATHEMATICS RESEARCH CENTER

610 WALNUT STREET
TELEPHONE (608) 263-2661

Dear Mr. Adams,

Please excuse my delay in answering your letter. I was in Hawaii for a month, and have just got to opening the collected mail.

I would be glad to read some (or all) of your chapters. You should, of course, address a similar request to Prof. S. C. Kleene, of the department of mathematics here at the University of Wisconsin. Are you acquainted with his paper "The theory of recursive functions, approaching its centennial", which appears on pp. 43–61 of the Bulletin (New Series) of the Amer. Math. Soc., vol. 5 (1981)? You should find it very helpful.

Sincerely

J. Barkley Rosser

Dear Mr. Adams,

Please excuse my delay in answering your letter. I was in Hawaii for a month, and have just got to opening the collected mail.

I would be glad to read some (or all) of your chapters. You should, of course, address a similar request to Prof. S. C. Kleene, of the department of mathematics here at the university of Wisconsin. Are you acquainted with his paper "The theory of recursive functions, approaching its centennial," which appears on pp. 43–61 of the *Amer. Math. Soc.*, vol. 5 (1981)? Your should find it very helpful.

Sincerely,

J. Barkley Rosser

B.10. Stephen C.Kleene to Author, September 9, 1982

Mathematics Department
University of Wisconsin—Madison

Van Vleck Hall
480 Lincoln Drive
Madison, Wisconsin 53706
Telephone: (608) 263-3053

uw madison

September 9, 1982

Mr. R. G. Adams
St. Albans College
29 Hatfield Road
St. Albans, Herts.
ENGLAND

Dear Mr. Adams:

Because of travels this summer (three weeks at the AMS 1982 Summer Institute in Recursion Theory at Cornell University beginning at the end of June, and four weeks in China, Japan and California beginning late in July), it is only now that I have got around to reading your Chapters 6 and 7, which you sent with your note of June 6.

There are some corrections I would like to make in my 1981 and 1981a) (see the bibliography at the end of your Chapter 6). I do not remember how many of them (if any) were marked in the reprints of them which I sent you in November.

1981

A. p.52, rt. col., line 5, add at the end the words "the first of". [This is to prevent what I say being taken to imply incorrectly that Gödel's second theorem was announced at the Königsberg conference.]

B. p.57, lt. col., line 21 from below, for "retain" read "contain". [The rewriting did not consist solely in removing other things from the original version of the thesis, but some new things were added.]

C. p.59, lt. cl., line 4 from below, add at the front the words "in 1934". [This is to make it explicit that what follows, all the way through the first two paragraphs of the rt. col., were in 1934, and not (except for Gödel's coming to Princeton) in the fall of 1933.]

D. p.60, lt. col., lines 17-18, for "Church (1936) and I (1936a) published equivalence proofs" to "I (1936a) published an equivalence proof". [When I wrote my paper for Annals of the History of Computing, I thought, without having checked carefully, that Church 1936 contained in outline an equivalence proof. Martin Davis brought it to my attention that it did not, and that indeed Church claimed to have proved only that every λ-definable function is recursive.]

Mr. R. G. Adams
Page 2
September 9, 1982

E. p.63, lt. col., line 4 from below, for "$\underline{1944}$" read "$\underline{1954}$".

F. p.64, lt. col., bottom line, for "Δ_0^1" read "Δ_1^0".

1981a)

G. p.43, Footnote 1, line 2, for "Richard A. Shore" read "Robert A. Soare".

    Corrections C and D have some relevance to your Chapter 6.

    I am not sure that your exposition would not convey the impression that Church (as well as I) had the proof of equivalence, whereas Church had only one of the two implications ($\underline{\lambda\text{-definable implies recursive}}$).

    I do not think I communicated my results in definite form to Church until my two papers 1936 and 1936a) were ready to be sent off around the end of June 1935. What Rosser may have communicated to Church I do not know.

    It may be noted that Church in his abstract in Bull. Amer. Math. Soc., vol. 41, pp. 332-333 (received March 22, 1935) mentions only general recursiveness and not λ-definability. This is plausibly explained by the supposition that at that time he did not know that $\underline{\text{recursive implies }\lambda\text{-definable}}$. Indeed, since he first proposed his thesis using λ-definability, it would be hard otherwise to explain his not including it in the March 1935 abstract.

    Your pp.22-23 suggests that Church "actually came out and asserted his thesis" $\underline{\text{after}}$ the general recursive functions had appeared. If by this you mean that he asserted it as his definite belief, in conversation with me (and probably others), $\underline{\text{that}}$ happened $\underline{\text{before}}$ Gödel had produced the notion of general recursive functions. I enclose a copy of Church's letter of November 29, 1935 (from which I quoted only briefly in 1981 p.59 rt. col.), which seems to make this clear.

    I can date the time Church definitely asserted his thesis to me (as I mention in "Reminiscences of Logicians") as in February or March 1934, as [came out] I only returned to Princeton (having been absent during the fall of 1933) on [with] February 7, 1934. It was later, in April or May 1934, that Gödel came out with his notion of general recursiveness. Thus your "Gödel arrived at Princeton and $\underline{\text{brought with}}$ $\underline{\text{him}}$ the notion of general recursive functions ..." (p.22, italics mine) is in error. Gödel arrived in Princeton in the fall of 1933.

    I am surprised by your statement on p.23 that "on April 19th, 1935 Church submitted to the publishers a preliminary copy of his paper 'An unsolvable problem ...' containing his thesis." April 19, 1935 is the date of Church's 10 minute talk ("preliminary report") delivered to a meeting of the A.M.S.; but I had not known of a preliminary copy (a manuscript) of the paper being submitted to the (eventual) publishers on that date. An $\underline{\text{abstract}}$ was ~~by BAMS~~ received on March 22, 1935. And, as I have stated, I am skeptical that the
(by B.A.M.S.)

Mr. R. G. Adams
Page 3
September 9, 1982

> paper was presumably in a late

<u>proof</u> of the equivalence was known at that time, though I am sure we were
<u>conjecturing</u> the equivalence.

On p.29, you mention that van Heijenoort gives "Über die Länge von Beweisen"
as a 1934 paper. Maybe van Heijenoort did not know that <u>Ergebnisse eines math.</u>
<u>Kolloquiums</u> has the date of Gödel's talk misprinted as "19 VI 1934" whereas it
was "19 VI 1935". This is clear from the dates of the preceding and following
colloquia. We don't know how soon it was after Church wrote me on July 6, 1935
"I have made a number of changes, since I saw you, in my paper "An unsolvable
problem of elementary number theory'", that he sent <u>that paper in for publication.</u>
It probably wasn't long, But, as Church's stage of composition by mid-June 1935,
there seems to me no evidence for giving Gödel the priority for the concept
<u>computable (reckonable)in a logic</u>. As it was used by both in papers written in
the middle of 1935, and Gödel was away from Princeton from May or June 1934 until
the <u>fall</u> of 1935, I doubt that Church and Gödel discussed it. For them to have
done so would put it in the first of the fourteen or so months that passed between
Gödel's bringing out the notion of general recursiveness and his presenting his
paper "Über die Länge von Beweisen" in Menger's colloquium. (And at least at the
beginning of this fourteen months, he didn't accept Church's thesis for general
recursiveness, as you know (p.34).) I think probably Church's November 29, 1935
letter covers the substance of his exchanges with Gödel in the academic year
1933-34, and thus before Gödel's return to Princeton in the fall of 1935. Thus
I think the conception of "reckonable" was conceived by Church and Gödel independently and about simultaneously.

P.1 bottom line. Do you want "characterizing" instead of "classifying", which
suggests subrecursive hierarchies?

No doubt you know some hyphens are missing (compare p.3 line 1 with p.13
line 5 f.b.). Also, some umlauts on Gödel.

Chapter 7

P.1 line 3. Was this not his bachelor's degree (undergraduate degree)? I
have the impression that Turing completed work for a Ph.D. in Princeton in 1938.
I am sure you have access to the facts.

P.2 line 13, I suggest adding "(real)" before "numbers" (this first place
only).

P.7, line 3. I think "decimal" is a misnomer when applied to an expansion
in 0's and 1's. (I would say a "dual expansion" rather than a "decimal expansion".)
However, Turing uses "decimal", but mends things a bit by saying "binary decimal"
(Davis 1965, p.118, line 8 from below). I suggest you add "binary" before "decimal"
at the cited place.

Mr. R. G. Adams
Page 4
September 9, 1982

    P.34, line 3, I suggest replacing "undertook" by "proposed to undertake".

                            Sincerely yours,

                            Stephen C. Kleene
                            Stephen C. Kleene

SCK/kjs
cc
encl.

P.S. References like "1981a)" tend to confuse because they violate the lemmas
on pairing of parentheses, Kleene 1952 p.24.

PPS. You will probably be interested by Martin Davis, "Why Gödel didn't have
Church's thesis", which covers some of the same territory you do in Chapter 6.
I have what I take to be a preprint of it, labeled "(revised December 26, 1981)",
but I do not know where it is to be published. Davis cites Judson C. Webb,
"Mechanism, Mentalism, and Metamathematics", Reidel, Dordrecht, Netherlands, 1980,
suggesting that it says something significant about Church's thesis; but I
haven't yet been able to see it.

Dear Mr. Adams:

Because of travels this summer (three weeks at the AMS 1982 Summer Institute in Recursion Theory at Cornell University beginning at the end of June, and four weeks in China, Japan and California beginning late in July), it is only now that I have got around to reading your Chapters 6 and 7, which you sent with your note of June 6.

There are some corrections I would like to make in my 1981 and 1981a) (see the bibliography at the end of your Chapter 6). I do not remember how many of them (if any) were marked in the reprints of them which I sent you in November.

1981

    A. p.52, rt. col., line 5, add at the end the words "the first of". [This is to prevent what I say being taken to imply incorrectly that Gödel's second theorem was announced at the Königsberg conference.]

    B. P.57, lt . col ., line 21 from below for "retain" read "contain". [The rewriting did not consist solely in removing other things from the original version of the thesis, but some new things were added.]

    C. p.59, lt. cl, line 4 from below, add at the front the words "in 1934". [This is to make it explicit that what follows, all the way through the first two paragraphs of the rt. col., were in 1934, and not (except for Gödel's coming to Princeton) in the fall of 1933.]

    D. p.60, lt. col., lines 17-18, for "Church (1936) and I (1936a) published equivalence proofs" to "I (1936a) published an equivalence proof". [When I wrote my paper for *Annals of the History of Computing*, I thought, without having checked carefully, that Church *1936* contained in outline an equivalence proof. Martin Davis brought it to my attention that it did not, and that indeed Church claimed to have proved only that every λ-definable function is recursive.]

E. p.63, lt. col., line 4 from below, for "1944" read "1954".

F. p.64, lt. col., bottom line, for "Δ_0^1" read "Δ_1^0". 1981a)

G. P.43, Footnote 1, line 2, for "Richard A. Shore" read "Robert A. Soare".

Corrections C and D have some relevance to your Chapter 6.

I am not sure that your exposition would not convey the impression that Church (as well as I) had the proof of equivalence, whereas Church had only one of the two implications (λ-*definable implies recursive*).

I do not think I communicated my results in definite form to Church until my two papers 1936 and 1936a) were ready to be sent off around the end of June 1935. What Rosser may have communicated to Church I do not know.

It may be noted that Church in his abstract in *Bull. Amer. Math. Soc.,* vol. 41, pp. 332–333 (received March 22, 1935) mentions only general recursiveness and not λ-definability. This is plausibly explained by the supposition that at that time he did not know that recursive implies λ-definable. Indeed, since he first proposed his thesis using λ-definability, it would be hard otherwise to explain his not including it in the March 1935 abstract.

Your pp.22–23 suggests that Church "actually came out and asserted his thesis" *after* the general recursive functions had appeared. If by this you mean that he asserted it as his definite belief, in conversation with me (and probably others), *that* happened *before* Gödel had produced the notion of general recursive functions. I enclose a copy of Church's letter of November 29, 1935 (from which I quoted only briefly in 1981 p.59 rt. col.), which seems to make this clear.

I can date the time Church definitely asserted his thesis to me (as I mention in "Reminiscences of Logicians") as in February or March 1934, as I only returned to Princeton (having been absent during the fall of 1933) on February 7,1934. It was later, in April or May 1934, that Gödel came out with his notion of general recursiveness. Thus your "Gödel arrived at Princeton and *brought with him* the notion of general recursive functions ..." (p. 22, italics mine) is in error. Gödel arrived in Princeton in the fall of 1933.

I am surprised by your statement on p.23 that "on April 19th, 1935 Church submitted to the publishers a preliminary copy of his paper 'An unsolvable problem ...' containing his thesis." April 19, 1935 is the date of Church's 10 minute talk ("preliminary report") delivered to a meeting of the A.M.S.; but I had not known of a preliminary copy (a manuscript) of the paper being submitted to the (eventual) publishers on that date. An *abstract* was received by B.A.M.S. on March 22, 1935. And, as I have stated, I am skeptical that the *proof* of the equivalence was known at that time, though I am sure we were *conjecturing* the equivalence.

On p.29, you mention that van Heijenoort gives "Über die Länge von Beweisen" as a 1934 paper. Maybe van Heijenoort did not know that *Ergebnisse eines math. Kolloquiums* has the date of Gödel' s talk misprinted as "19 VI 1934" whereas it was "19 VI 1935". This is clear from the dates of the preceding and following colloquia. We don't know how soon it was after Church wrote me on July 6, 1935 "I have made a number of changes, since I saw you, in my paper 'An unsolvable problem of elementary number theory"', that he sent that paper in for publication. It probably wasn't long. But, as Church's paper was presumably in a late stage of composition by mid-June 1935, there seems to me no evidence for giving Gödel the priority for the concept *computable (reckonable)in a logic*. As it was used by both in papers written in the middle of 1935, and Gödel was away from Princeton from May or June 1934 until the *fall* of 1935, I doubt that Church and Gödel discussed it. For them to have done so would put it in the first of the fourteen or so months that passed between Gödel's bringing out the notion of general recursiveness and his presenting his paper "Über die Länge von Beweisen" in Menger's colloquium. (And at least at the beginning of this fourteen months, he didn't accept Church's thesis for general recursiveness, as you know (p.34).) I think probably Church's November 29, 1935 letter covers the substance of his exchanges with Gödel in the academic year 1933–34, and thus before Gödel's return to Princeton in the fall of 1935. Thus I think the conception of "reckonable" was conceived by Church and Gödel independently and about simultaneously.

P.1 bottom line. Do you want "characterizing" instead of "classifying", which suggests subrecursive hierarchies?

No doubt you know some hyphens are missing (compare p.3 line 1 with p.13 line 5 f. b.). Also, some umlauts on Gödel.

Chapter 7

P.1 line 3. Was this not his bachelor's degree (undergraduate degree)? I have the impression that Turing completed work for a Ph.D. in Princeton in 1938. I am sure you have access to the facts.

P.2 line 13, I suggest adding "(real)" before "numbers" (this first place only).

P.7, line 3. I think "decimal" is a misnomer when applied to an expansion in 0's and 1's. (I would say a "dual expansion" rather than a "decimal expansion".) However, Turing uses "decimal", but mends things a bit by saying "binary decimal" (Davis 1965, p.118, line 8 from below). I suggest you add "binary" before "decimal" at the cited place.

P. 34, line 3, I suggest replacing "undertook" by "proposed to undertake".

Sincerely yours,

Stephen C. Kleene

SCK/kjs

cc

encl.

P.S. References like "1981a)" tend to confuse because they violate the lemmas on pairing of parentheses, Kleene 1952 p.24.

PPS. You will probably be interested by Martin Davis, "Why Gödel didn't have Church's thesis", which covers some of the same territory you do in Chapter 6. I have what I take to be a preprint of it, labeled "(revised December 26, 1981)"; but I do not know where it is to be published. Davis cites Judson C. Webb, *Mechanism, Mentalism, and Metamathematics,* Reidel, Dordrecht, Netherlands, 1980, suggesting that it says something significant about 'Church's thesis; but I haven't yet been able to see it.

B.11. Stephen C. Kleene to Author, April 4, 1983

Mr. R. G. Adams,
St. Albans College,
29 Hatfield Road,
St. Albans, Herts.,
ENGLAND

Van Vleck Hall,
Univ. of Wisconsin,
Madison, WI 53706, U.S.A.,
April 4, 1983

Dear Mr. Adams:

In reply to your letter of 13 March 1983, it is hard to be quite positive about whether Church knew that general recursiveness implies lambda-definability at the time he wrote his abstract (received March 22, 1935). If we take him at his word in Footnote 16 of the actual paper, and if it is the fact (as I believe I recall correctly) that I didn't have my proof in final form till June 1935, when I put copies of my two papers which appeared in 1936, then he did not! in his hands

How then explain the fact (which you have just called to my attention) that the last sentence of the abstract seems to contradict this, unless we can come up with a proof of it not using rec. → λ-def. (and I don't)? (The abstract says nothing about well-formed being in the modified sense of end Footnote 16.)

I think all of us at Princeton anticipated that the equivalence of λ-definability and general recursiveness would be proved.

Is it possible that there was some confusion or vacillation on the part of Church in the early part of 1935 as to what he believed at the moment to be definitely established? Only very recently there came to my eyes a letter of Church to Bernays dated 23 January 1935 in which, after defining what it means for a formula F in his λ-notation (as we now call it) to represent an intuitively defined function, Church says, "... it can be proved that a formula can be found to represent any function which is recursive [in the sense introduced in § 9 of Gödel's 1934 lectures] (the proof is due to Rosser)." This of course contradicts Footnote 16 (as there is nothing there about the modified form of well-formed being used).

I have wondered what the proof by Rosser alluded to in Footnote 16 consisted in. It seemed to me that the proof of the result in its full generality (with systems of equations for general recursiveness allowed which may bear no resemblance to the sorts of recursions previously considered) requires the sort of analysis I gave in establishing my normal form theorem (Math. Ann., 1936 p. 736 IV).

In Footnote 16 Church says the implication rec. → λ-def "can be proved as a straightforward application of the methods introduced by Kleene in [his 1935 paper]." Yes, that is true, once one has the normal form theorem. I have to wonder whether Rosser's proof (with the modified sense of λ-definable) was quite decisive. I never got any reaction, when I announced my normal form theorem, to the effect that Rosser (or Church) had been thinking along those lines. paper? and to give it top billing in stating his thesis in the

Did I mention to you earlier another possible factor affecting Church's choice in the abstract to use general recursiveness rather than λ-definability. This is that general recursiveness was thought to have more sex appeal to mathematicians - to be more in the line of a familiar mathematical development, going back to Dedekind 1888, Peano 1891, Skolem 1923, Hilbert 1926 (with Ackermann 1928), etc. In contrast, λ-definability was a rather new idea. Compare my A.H.C. article of Jan. 1981 (I sent you a reprint), p. 62 lt. col. 3rd paragraph.

In a letter to me of November 15, 1935, Church wrote, "Both Gödel and Bernays are very much interested in your result that all general recursive functions can by obtained from primitive recursive functions by means of the epsilon operator [my normal form theorem]. It is, of course, natural that they should be quicker to see the importance of a result like this than of one concerning lambda-definition or the like."

Sincerely,

Stephen C. Kleene

Dear Mr. Adams:

In reply to your letter of 13 March 1983, it is hard to be quite positive about whether Church knew that *general recursiveness implies lambda-definability* at the time he wrote his abstract (received March 22, 1935). If we take him at his word on Footnote 16 of the actual paper, and if it is the fact (as I believe I recall correctly) that I didn't have my proof in final form till June 1935, when I put in his hands copies of my two papers which appeared in 1936, then he did not!

How then explain the fact (which you have just called to my attention) that the last sentence of the abstract seems to contradict this, unless we can come up with a proof of it not using *rec.* $\rightarrow$ λ-*def.* (and I don't)? (The abstract says nothing about *well-formed* being in the modified sense of end Footnote 16.)

I think all of us at Princeton anticipated that the equivalence of λ-definability and general recursiveness *would be proved*.

It is possible that there was some confusion or vacillation on the part of Church in the early part of 1935 as to what he believed at the moment to be definitely established? Only very recently there came to my eyes a letter of Church to Bernays dates 23 January 1935 in which, after defining what it means for a formula F in his λ-notation (as we now call it) to represent an intuitively defined function, Church says, "... it can be proved that a formula can be found to represent any function which is recursive [in the sense introduced in § 9 of Gödel's 1934 lectures] (the proof is due to Rosser)." This of course contradicts Footnote 16 (as there is nothing there about the modified form of well-formed being used).

I have wondered what the proof by Rosser alluded to in Footnote 16 consisted in. It seemed to me that the proof of the result in its full generality (with systems of equations for general recursiveness allowed which may bear no resemblance to the sorts of recursions previously considered) requires the sort of analysis I gave in establishing my normal for theorem (*Math. Ann.*, 1936 p. 736 IV).

In Footnote 16 Church says the implication *rec.* $\rightarrow$ λ-*def.* "can be proved as a straightforward application of the methods introduced by Kleene in [his 1935 paper]." Yes, that is true, once one has the normal form theorem. I have to wonder whether Rosser's proof (with the modified sense of λ-definable) was quite decisive. I never got any reaction, when I announced my normal for theorem, to the effect that Rosser (or Church) had been thinking along those lines.

Did I mention to you earlier another possible factor affecting Church's choice in the abstract to use general recursiveness rather than λ-definability, and to give it top billing in stating his thesis in the paper? This is that general recursiveness was thought to have more sex appeal to mathematicians – to be more in line of a familiar mathematical development, going back to Dedekind 1888, Peano 1891, Skolem 1923, Hilbert 1926 (with Ackermann 1928), etc. In contrast, λ-definability was a rather new idea. Compare my *A.H.C.* article of Jan. 1981 (I sent you a reprint), p. 62 lt. col. 3rd paragraph.

In a letter to me of November 15, 1935, Church wrote "Both Gödel and Bernays are very much interested in your result that all general recursive functions can be obtained from primitive recursive functions by means of the epsilon operator [my normal form theorem]. It is, of course, natural that they should be quicker to see the importance of a result like this than of one concerning lambda-definition of the like."

Sincerely yours,

Stephen C. Kleene

Bibliography

[1] Wilhelm Ackermann, Begründung des "tertium non datur" mittels der Hilbertschen Theorie der Widerspruchsfreiheit, Math. Ann. **99** (1924), 1–36.

[2] ———, Zum Hilbertschen Aufbau der reelen Zahlen, Math. Ann. **93** (1928), 118–133.

[3] Anonymous, Unsigned note on Herbrand, Annales de l'Université de Paris **6** (1931), 186–189.

[4] Paul Benacerraf and Hilary Putnam (eds.), Philosophy of Mathematics, Prentice Hall, New Jersey, 1964.

[5] Paul Bernays, Zusatz zu Hilberts Vortrag über 'Die Grundlagen der Mathematik, Abhandlungen aus dem mathematischen Seminar der Hamburgischen Universität **6** (1927), 89–92.

[6] ———, David Hilbert, Encyclopedia of Philosophy, vol. 3, Macmillian, New York, 1967, pp. 89–92.

[7] Evert Willem Beth, The Foundations of Mathematics, North-Holland, Amsterdam, 1959.

[8] ———, Mathematical Thought, D. Reidel, Dordrecht, 1965.

[9] Luitzen Eghertus Jan Brouwer, Über die Bedeutung des Satzes vom ausgeschlossenen Dritten in der Mathematik, insbesondere in der Funktionentheorie, J. Reine Angew. Math. **154** (1923), 1–7.

[10] ———, Intuitionistische Betrachtungen über den Formalismus, Koninklijke Akademie van Wetenschappen te Amsterdam, Proceedings of the Section of Sciences **31** (1927), 374–379.

[11] Cesare Burali-Forti, Una questione sui numeri transfiniti, Rendiconti del Circolo matematico di Palermo **11** (1897), 154–164.

[12] W. H. Bussey, The Origin of Mathematical Induction, Amer. Math. Monthly **24** (1917), 199–207.

[13] Cristian Calude and Solomon Marcus, The First Example of a Recursive Function which is not Primitive Recursive, Historia Mathematica **6** (1979), 380–384.

[14] Georg Cantor, Über die Ausdehnung eines Satzes aus der Theorie der trigonometrischen Reihen, Math. Ann. **5** (1872), 123–132.

[15] ———, Über unendliche, lineare Punktmannichfaltigkeiten, Math. Ann. **21** (1883), 545–591.

[16] Rudolf Carnap, The Logical Syntax of Language, Harcourt and Brace, New York, 1937.

[17] Alonzo Church, A Set of Postulates for the Foundations of Logic, Annals of Mathematics **33** (1932), 346–366.

[18] ———, A Set of Postulates for the Foundations of Logic (Second Paper), Annals of Mathematics **34** (1932), 839–864.

[19] ———, The Richard Paradox, Amer. Math. Monthly **41** (1934), 356–361.

[20] ———, A Proof of Freedom from Contradiction, Proc. Nat. Acad. Sciences **21** (1935), 275–281.

[21] ———, Correction: A Note on the Entscheidungsproblem, J. Symbolic Logic **1** (1936), 101–102.

[22] ———, A Note on the Entscheidungsproblem, J. Symbolic Logic **1** (1936), 40–41.

[23] ———, An Unsolvable Problem of Elementary Number Theory, Amer. J. Math. **58** (1936), 345–363.

[24] ———, Review of Turing 1936, J. Symbolic Logic **2** (1937), 42–43.

[25] ———, The Constructive Second Number Class, Bull. Amer. Math. Soc. **44** (1938), 224–238.

[26] ———, Introduction to Mathematical Logic, Princeton University Press, Princeton, 1956.

[27] Alonzo Church and Stephen Cole Kleene, Formal Definitions in the Theory of Ordinal Numbers, Fundamenta Mathematicæ **28** (1936), 11–21.

[28] Alonzo Church and John Barkley Rosser, Some Properties of Conversion, Trans. Amer. Math. Soc. **39** (1936), 472–482.

[29] John Crossley (ed.), Reminiscences of Logicians. Algebra and Logic. Papers from the 1974 Summer Research Institute of the Australian Mathematical Society at Monash University, Australian Mathematical Society, 1975.

[30] Haskell Brooks Curry, An Analysis of Logical Substitution, Amer. J. Math. **51** (1929), 363–384.

[31] ———, Grundlagen der Kombinatorischen Logik, Amer. J. Math. **52** (1930), 509–536, 789–834.

[32] ———, Some Additions to the Theory of Combinators, Amer. J. Math. **54** (1932), 551–558.

[33] ———, The Paradox of Kleene and Rosser, Trans. Amer. Math. Soc. **50** (1941), 454–516.

[34] Martin Davis, Computability and Unsolvability, McGraw-Hill, New York, 1958.

[35] Martin Davis (ed.), The Undecidable, Raven Press, New York, 1965.

[36] ———, Recursive Function Theory, Encyclopedia of Philosophy, vol. 7, Macmillian, New York, 1967, pp. 89–95.

[37] Richard Dedekind, Stetigkeit und Irrationale Zahlen, Braunschweig, Friedrich Beiweg and Son, 1872.

[38] ———, Was Sind und Was Sollen die Zahlen?, Friedrich Beiweg and Son, Braunschweig, 1888.

[39] ———, Essays on the Theory of Numbers, Open Court, Chicago, 1901, Translation by Wooster Woodruff Beman.

[40] ———, Letter to Keferstein, 27 February 1890, From Frege to Gödel: A Source Book in Mathematical Logic, 1879–1931 (Jean van Heijenoort, ed.), Harvard University Press, Cambridge, MA, 1967.

[41] Howard DeLong, A Profile of Mathematical Logic, Addison-Wesley, Reading, Massachusetts, 1970.

[42] Abraham A. Fraenkel, Der Begriff 'definit" und die Unabhängigkeit des Auswahlsaxioms, Sitzungsberichte der Preussischen Akademie der Wissenschaften, Physikalisch-mathematische Klasse (1922), 253–257.

[43] Abraham A. Fraenkel, Yehoshua Bar-Hillel, and Azriel Levy, Foundations in Set Theory, 2 ed., North-Holland, Amsterdam, 1973.

[44] Friedrich Ludwig Gottlob Frege, Begriffsschrift, eine der arithmetischen nachgebildete Formelsprache des reinen Denkens, Halle, 1879.

[45] ———, Die Grundlagen der Arithmetik, eine logisch-mathematische Untersuchang über den Begriff der Zahl, Breslau, 1884.

[46] ———, On the Foundations of Geometry and Formal Theories of Arithmetic, Yale University Press, New Haven and London, 1971.

[47] Hans Freudenthal, Hilbert, David, Dictionary of Scientific Biography (Charles Coulston Gillispie, ed.), Charles Scribner's and Sons, 1972, pp. 388–395.

[48] Gerhard Gentzen, Die Widerspruchsfreiheit der reinen Zahlentheorie, Math. Ann. **112** (1936), 493–565.

[49] Kurt Gödel, Die Vollständigkeit der Axiome des logischen Funktionenkalküls, Monatshefte für Mathematik und Physik **37** (1930), 349–360.

[50] ———, Einige metamathematische Resultate über Entscheidungsdefinitheit und Widerspruchsfreiheit, Anzeiger der Akademie der Wissenschaften in Wein, Mathematisch-naturwissenschaftliche Klasse **67** (1930), 214–215.

[51] ———, Über formal unentscheidbare Sätze der Principia mathematica und venwandter Systeme, *I*, Monatshefte für Mathematik und Physik **38** (1931), 173–198.

[52] ———, Über Vollständigkeit und Widerspruchsfreiheit, Ergebnisse eines mathematischen Kolloquiums **3** (1931), 12–13.

[53] ———, Remarks contributed to a Diskussion zue Grundlegung der Mathematik, Erkenntnis **2** (1932), 147–148.

[54] ———, On Undecidable Propositions of Formal Mathematical Systems, Lecture Notes (S.C. Kleene and J.B. Rosser, eds.), The Institute for Advanced Studies, Princeton, 1934, pp. 41–74.

[55] ———, Über die Länge von Beweisen, Ergebnisse eines mathematischen Kolloquiums **7** (1936), 23–24.

[56] ———, The Consistency of the Axiom of Choice and of the Generalised Continuum-Hypothesis, Proc. Nat. Acad. Sciences **24** (1938), 556–557.

[57] ———, The Consistency of the Axiom of Choice and of the Generalised Continuum-Hypothesis with the Axioms of Set Theory, Lecture Notes (G.W. Brown, ed.), The Institute for Advanced Studies, Princeton, 1940.

[58] W. D. Goldfarb (ed.), Logical Writings, Reidel, Dordrecht, 1971.

[59] Reuben Louis Goodstein, Recursive Number Theory, North-Holland, Amsterdam, 1957.

[60] ———, Recursive Analysis, North-Holland, Amsterdam, 1962.

[61] ———, Development of Mathematical Logic, Logos, London, 1971.

[62] Herman Grassmann, Lehrbuch der Arithmetik für höhere Lehranstalten, Th Fr. Enslin, Berlin, 1861.

[63] C. Hartshorne and P. Weiss (eds.), Collected Papers of Charles Sanders Peirce, vol. 3, Harvard University Press, Cambridge, Massachusetts, 1933.

[64] Jean van Heijenoort, From Frege to Gödel – A Source Book in Mathematical Logic, 1879–1931, Harvard University Press, Cambridge, Massachusetts, 1967.

[65] ———, Gödel's Theorem, Encyclopedia of Philosophy, vol. 3, Macmillian, New York, 1967, pp. 348–357.

[66] Jacques Herbrand, Les Bases de la Logique Hilbertienne, Revue de métaphysique et de morale **37** (1930), 243–255.

[67] ———, Sur la non-contradiction de l'arithmétique, J. Reine Angew. Math. **166** (1931), 1–8.

[68] David Hilbert, Grundlagen der Geometrie, Teubner, Leipzig, 1899.

[69] ———, Mathematische Probleme: Vortrag, gehalten auf dem internationalen Mathematiker-Kongress zu Paris 1900, Nachrichten von der Königlichen Gesellschaft der Wissenschaften zu Göttingen (1900), 253–297.

[70] ———, Über den Zahlbegriff, Jahresbericht der Deutschen Mathematiker-Vereinigung **8** (1900), 180–194.

[71] ———, Mathematische Probleme, Trans. Amer. Math. Soc. **8** (1902), 437–479.

[72] ———, Über die Grundlagen der logik und der Arithmetik, pp. 174–185, B. G. Teubner, Leipzig, 1904.

[73] ———, Axiomatisches Denke, Math. Ann. **78** (1917), 405–415.

[74] ———, Die Logischen Grundlagen der Mathematik, Math. Ann. **88** (1922), 151–165.

[75] ———, Neubegründung der Mathematik. Erste Mitteilung, Abhandlungen aus dem mathematischen Seminar der Hamburgischen Universität **1** (1922), 157–177.

[76] ———, Über das Unendliche, Math. Ann. **95** (1925), 161–190.

[77] ———, Die Grundlagen der Mathematik, Abhandlungen aus dem mathematischen Seminar der Hamburgischen Universität **6** (1927), 65–85.

[78] ———, Gesammelte Abhandlungen, vol. 3, Springer-Verlag, Berlin, 1935.

[79] David Hilbert and Wilhelm Ackermann, Grundzüge der theoretischen Logik, Springer-Verlag, Berlin, 1928.

[80] David Hilbert and Paul Bernays, Grundlagen der Mathamtik, vol. 1, Springer-Verlag, Berlin, 1934.

[81] ———, Grundlagen der Mathamtik, vol. 2, Springer-Verlag, Berlin, 1939.

[82] Jill Humphries, Gödel's Proof and the Liar Paradox, Notre Dame Journal of Fomal Logic **120** (1979), 535–544.

[83] László Kalmár, An Argument Against the Plausibility of Church's Thesis. Constructivity in Mathemtics, Proceedings of the Colloquium held at Amsterdam, 1957, North-Holland, Amsterdam, 1959, pp. 72–80.

[84] Stephen Cole Kleene, Proof by Cases in Formal Logic, Annals of Mathematics **35** (1934), 529–544.

[85] ———, A Theory of Positive Integers in Formal Logic, Amer. J. Math. **57** (1935), 153–173, 219–244.

[86] ———, General Recursive Functions of Natural Numbers, Math. Ann. **112** (1936), 727–742.

[87] ———, λ-Definability and Recursiveness, Duke Mathematical Journal **2** (1936), 340–353.

[88] ———, A Note on Recursive Functions, Bull. Amer. Math. Soc. **42** (1936), 544–546.

[89] ———, On Notation for Ordinal Numbers, J. Symbolic Logic **3** (1938), 150–155.

[90] ———, Recursive Predicates and Quantifiers, Trans. Amer. Math. Soc. **53** (1943), 41–73.

[91] ———, Introduction to Metamathematics, North-Holland, Amsterdam, 1952.

[92] ———, Mathematical Logic, John Wiley, New York, 1967.

[93] ———, The Work of Kurt Gödel, J. Symbolic Logic **41** (1976), 761–778.

[94] ———, Origins of Recursive Function Theory, Ann. Hist. Computing **3** (1981), 52–67.

[95] ———, The Theory of Recursive Functions, Approaching its Centennial, Bull. Amer. Math. Soc. **5** (1981), 43–61.

[96] Stephen Cole Kleene and John Barkley Rosser, The Inconsistency of Certain Formal Logics, Annals of Mathematics **36** (1935), 630–636.

[97] Morris Kline, Mathematical Thought from Ancient to Modern Times, Oxford University Press, New York, 1972.

[98] Greoffrey T. Kneebone, Mathematical Logic and the Foundations of Mathematics, Van Nostrand, London, 1963.

[99] Julius König, Neue Grundlagen der Logik, Arithmetik und Mengenlehre, Veit, Leipzig, 1914.

[100] Georg Kreisel, Hilbert's Programme, Dialectica **12** (1958), 346–372.

[101] Paul Lévy, Remarques sur un théorème de Paul Cohen, Revue de métaphysique et de morale **69** (1964), 88–94.

[102] Leopold Löwenheim, Über Möglichkeiten im Relativkalkül, Math. Ann. **76** (1915), 447–470.

[103] Andrey Andreyevich Markov, On the Impossibility of Certain Algorithms in the Theory of Associative Systems, Comptes rendus (Doklady) de l'Academie des Sciences de l'USSR **55** (1947), 583–586.

[104] Kenneth Ownsworth May, Bibliography and Research Manual of History of Mathematics, Toronto Press, Toronto, 1973.

[105] Elliott Mendelson, On Some Recent Criiticism of Church's Thesis, Notre Dame Journal of Formal Logic **4** (1963), 201–205.

[106] _____, Introduction to Mathematical Logic, Van Nostrand, Princeton, 1964.

[107] Andrzej Mostowski, Thirty Years of Foundational Studies: Lectures on the Development of Mathematical Logic and the Study of the Foundations of Mathematics in 1930-1961, Blackwell, Oxford, 1966.

[108] Alan Musgrave, Logicism Revisited, British Journal for the Philosophy of Science **28** (1977), 99–127.

[109] Ernest Nagel and James R. Newman, Gödel's Proof, New York University Press, New York, 1958.

[110] John von Neumann, Eine Axiomatisierung der Mengenlehre, J. Reine Angew. Math. **154** (1925), 219–240.

[111] _____, Zur Hilbertschen Beweistheorie, Mathematische Zeitschrift **26** (1927), 1–46.

[112] _____, Collected Works, Pergamon Press, New York, 1961.

[113] M. H. A. Newman, Alan Mathison Turing, Biographical Memoirs of the Fellows of the Royal Society **1** (1955), 253–263.

[114] Charles Parsons, Mathematics, Foundations of, Encyclopedia of Philosophy, vol. 5, Macmillian, New York, 1967, pp. 188–213.

[115] Moritz Pasch, Vorlesungen Über neuere Geometrie, Teubner, Leipzig, 1882.

[116] Giuseppe Peano, Arithmetices principia, nova methodo exposita, Bocca, Turin, 1889.

[117] Charles Sanders Peirce, On the Logic of Number, Amer. J. Math. **4** (1881), 85–95.

[118] Rózsa Péter, Rekursive Funktionen, Verhandlungen des Internationalen Mathematiker-Kongresses Zürich 1932, vol. 2, 1932, pp. 336–337.

[119] _____, Über den Zusammenhang der verschiedenen Begriffe der rekursiven Funktionen, Math. Ann. **110** (1934), 612–632.

[120] _____, Konstruktion nichtrekursiver Funktionen, Math. Ann. **111** (1935), 42–60.

[121] _____, Über die mehrfache Rekursion, Math. Ann. **113** (1936), 489–527.

[122] _____, Rekursive Funktionen, Akademischer Verlag, Budapest, 1951.

[123] _____, Rekursivität und Konstruktivität, Constructivity in Mathematics: Proceedings of the Colloquium held at Amsterdam, 1957 (Amsterdam), North-Holland, 1959, pp. 226–233.

[124] Henri Poincaré, Les mathématiques et la logique, Revue de metaphysique et de morale **14** (1905–6), 17–34, 294–317.

[125] _____, Les mathématiques et la logique, Revue de metaphysique et de morale **13** (1905–6), 815–835.

[126] _____, Science et méthode, Flamarion, Paris, 1908.

[127] Emil Leon Post, Introduction to a General Theory of Elementary Propositions, Amer. J. Math. **43** (1921), 163–185.

[128] _____, Finite Combinatory Processes, Formulation 1, J. Symbolic Logic **1** (1936), 103–105.

[129] _____, Recursively Enumerable Sets of Positive Integers and Their Decision Problems, Bull. Amer. Math. Soc. **50** (1944), 284–316.

[130] _____, Recursive Unsolvability of a Problem of Thue, J. Symbolic Logic **12** (1947), 1–11.

[131] Mojzesz Presburger, Über die Vollständigkeit eines gewissen Systems der Arithmetik ganzer Zahlen, in welchem die Addition als einzige Operation hervortritt, Sprawozdanie z I Kongresu matematyków krajów slowianskich, Warszawa 1929, 1930, pp. 92–101, 395.

[132] Constance Reid, Hilbert, Allen and Unwin, London, 1970.

[133] Raphael Mitchel Robinson, Primitive Recursive Functions, Bull. Amer. Math. Soc. **53** (1947), 925–942.

[134] Robert Rogers, Mathematical Logic and Formalised Theories, North-Holland, Amsterdam, 1971.

[135] Hartley Rogers Jr., Theory of Recursive Functions and Effective Computability, McGraw-Hill, New York, 1967.

[136] John Barkley Rosser, A Mathematical Logic without Variables, Annals of Mathematics **36** (1935), 127–150.

[137] _____, A Mathematical Logic without Variables, Duke Mathematical Journal **1** (1935), 328–355.

[138] _____, Extension of some Theorems of Gödel and Church, J. Symbolic Logic **1** (1936), 87–91.

[139] _____, Review of Gödel 1936, J. Symbolic Logic **1** (1936), 116.

[140] _____, An Informal Exposition of Proofs of Gödel's Theorem and Church's Theorem, J. Symbolic Logic **4** (1939), 53–60.

[141] Bertrand Russell, Les Paradoxes de la Logique, Revue de métaphysique et de moral **14** (1906), 627–650.

[142] _____, Mathematical Logic as Based on the Theory of Types, Amer. J. Math. **30** (1908), 222–262.

[143] _____, La Théorie des Types Logiques, Revue de métaphysique et de moral **18** (1910), 263–301.

[144] _____, Introduction to Mathematical Philosophy, Allen and Unwin, London, 1919.

[145] Moses Schönfinkel, Über die Bausteine der mathematischen Logik, Math. Ann. **92** (1924), 305–316.

[146] Ernst Schröder, Vorlesungen über die Algebra der Logik (Exakte Logik), vol. 1, Teubner, Leipzig, 1890.

[147] _____, Vorlesungen über die Algebra der Logik (Exakte Logik), vol. 2, Part 1, Teubner, Leipzig, 1891.

[148] _____, Vorlesungen über die Algebra der Logik (Exakte Logik), vol. 3, Teubner, Leipzig, 1895.

[149] _____, Vorlesungen über die Algebra der Logik (Exakte Logik), vol. 2, Part 2, Teubner, Leipzig, 1905.

[150] Joseph R. Shoenfield, Mathematical Logic, Addison-Wesley, Reading, Massachusetts, 1967.

[151] Thoralf Skolem, Logisch-kombinatorische Untersuchungen über die Erfüllbarkeit oder Beweisbarkeit mathematischer Sätze nebst einem Theoreme über dichte Mengen, Skriftner utgit ar Videnskapsselskaper i Kristiania **4** (1920), 4–36.

[152] _____, Einige Bemerkungen zur axiomtischen Begründung der Mengenlehre, Matematikerkongressen i Helsingfors den 4-7 Juli 1922, Den femte skandinaviska matematikerkongressen, Redogörelse (1922), 217–232.

[153] _____, Begründung der elementaren Arithmetik durch die rekurrierende Denkweise ohne Anwendungscheinbarer Veränderlichen mit unendlichem Ausdehnungsbereich, Videnskapsselskapets skrifter, I Matematisk-naturvidenskabelig klasse, No. 6 (1923), 1–38.

[154] _____, Über die Grundlagendiskussionen in der Mathematik, Den syvende skandinaviske matematikerkrongress i Oslo 19-22 August 1929, 1929, pp. 3–21.

[155] Gabriel Sudan, Sur le nombre transfini ω^ω, Bulletin mathématique de la Société roumaine des Sciences **30** (1927), 11–30.

[156] M. E. Szabo (ed.), The Collected Papers of Gerhard Gentzen, North-Holland, Amsterdam, 1969.

[157] Alfred Tarski, Einige Betrachtungen über die Begriffe der ω-Widerspruchsfreiheit und der ω-Vollständigkeit, Monatshefte für Mathematik und Physik **40** (1933), 97–112.

[158] ———, Der Wahrheitsbegriff in den formalisierten Sprachen, Studia Philosophica **1** (1936), 261–405.

[159] ———, Logic, semantics, metamathematics, Oxford University Press, Oxford, 1956.

[160] Alan Mathison Turing, On Computable Numbers, with an Application to the Entscheidungsproblem, Proc. Lond. Math. Soc. **42** (1936), 230–265.

[161] ———, Computability and λ-Definability, J. Symbolic Logic **2** (1937), 153–163.

[162] ———, A Correction: On Computable Numbers, with an Application to the Entscheidungsproblem, Proc. Lond. Math. Soc. **43** (1937), 544–546.

[163] ———, The $\mathfrak{p}$-Punction in λ-K-Conversion, J. Symbolic Logic **2** (1937), 164.

[164] Sara Turing, Alan M. Turing, Heffer, Cambridge, 1959.

[165] Hao Wang, The Axiomatisation of Arithmetic, J. Symbolic Logic **22** (1957), 145–158.

[166] ———, Logic, Computers and Sets, Chelsea, New York, 1970.

[167] ———, From Mathematics to Philosophy, Routledge, London, 1974.

[168] ———, Kurt Gödel's Intellectual Dervelopment, The Mathematical Intelligencer **1** (1978), 182–185.

[169] ———, Some Facts about Kurt Gödel, J. Symbolic Logic **46** (1981), 653–659.

[170] Hermann Weyl, Über die Definitionen der mathematischen Grundbegriffe, Mathematisch-naturwissenschaftliche Blätter **7** (1910), 93–95, 109–113.

[171] ———, Das Kontinuum. Kritische Untersuchungen über die Grundlagen der Analysis, Veit, Leipzig, 1918.

[172] ———, Diskussionsbemerkungen zu dem zweiten Hilbertschen Vortrag über die Grundlagen der Mathematik, Abhandlungen aus dein Mathematischen Seminar der Hamburgischen Universität **6** (1927), 86–88.

[173] ———, David Hilbert and his Mathematical Work, Bull. Amer. Math. Soc. **50** (1944), 612–654.

[174] Alfred North Whitehead and Bertrand Russell, Principia Mathematica, vol. 1, Cambridge University Press, Cambridge, 1910.

[175] ———, Principia Mathematica, vol. 2, Cambridge University Press, Cambridge, 1912.

[176] ———, Principia Mathematica, vol. 3, Cambridge University Press, Cambridge, 1913.

[177] Ernst Zermelo, Untersuchungen über die Grundlagen der Mengenlehre 1, Math. Ann. **65** (1908), 261–281.

www.ingramcontent.com/pod-product-compliance
Lightning Source LLC
Chambersburg PA
CBHW060328200326
41519CB00011BA/1866